LOWELL TARLING

with assistance from

PETER REES, EXECUTIVE PRODUCER, MYTHBUSTERS

VACUUM PUMP
PRESSURE BOX

NOT TO SCALE
LAWYER FRAME FOR CORRECT MASS DIST.
STERS

Published by Wilkinson Publishing Pty Ltd
ACN 006 042 173
2 Collins St
Melbourne Vic 3000
Ph: (03) 9654 5446
www.wilkinsonpublishing.com.au

First published 2006

Readers should not engage in the activities described in this book. Readers performing the activities in this book do so at their own risk and descretion. Every effort has been made to ensure this book is free from error or omission. However, Beyond Entertainment, Wilkinson Publishing, the Author, the Editor or their respective employees or agents, do not warrant or endorse the activities described in this book and specifically disclaim any responsibility for injury, loss, or damage occasioned to any person acting on or refraining from action as a result of the material in this book whether or not such injury, loss or damage is in any way due to any negligent act or omission, breach of duty or default on the part of Beyond Entertainment, Wilkinson Publishing, the Author, the Editor or their respective employees or agents.

National Libraries of Australia
Cataloguing-in-Publication data:

Tarling, Lowell.
Busted : 17 mythbusters that could happen to you.

ISBN 1 921203 10 2.

1. Mythbusters (Television program). 2. Mythology. 3. Urban folklore. 4. Science - Experiments. I. Title.

398.2

Page and cover design by Spike Creative Pty Ltd.
Prahran, Victoria. Ph: (03) 9525 0900.
www.spikecreative.com.au

Printed in Singapore.

About the Author

Lowell Tarling has been a professional writer since 1980. He has written three novels, 15 non-fiction books, and has ghosted 60 or so books which have appeared in other people's names.

Lowell is the former editor of the Dynamic Small Business (DSB) and Franchising magazines. Newsletter editorships include The Small Business Letter, The Leadership Letter and Time Management.

His best works, on singer Tiny Tim and artist Martin Sharp (ongoing since 1982), to date remain unpublished.

Website: www.lowelltarling.com.au

Warning! Read This!

DO NOT attempt these experiments, tests, trials, or any activity in this book at home, work, or anywhere else for that matter. The Mythbusters Team is trained, experienced and has resources which include safety and medical teams.

BUSTED
CONTENTS:

PREFACE: *Who are we?* 8

INTRO: *The Pioneer* 14

1. **Light your Tongue** 28
A man was hit by a bolt of lightning. Was the lightning attracted to his metal tongue stud?

2. **Is this a Bust?** 34
When you go up in an airplane the reduction in pressure will explode your implants.

3. **A Clean Kill** 42
Water conducts electricity to the point where an electrical appliance dropped into a bath will electrocute, and probably kill, the person in it.

4. **Think Quick** 52
We know that quicksand exists, but can it pull you down and kill you?

5. **Going Down?** 62
You're in an elevator, the cables are cut, you're plummeting rapidly towards the bottom of the elevator shaft and right at the very last second you jump as hard as you can – and that saves your life.

6. **A Tissue Issue** 72
A tissue box sitting on the back shelf of your car could gain enough velocity in a car crash to kill you.

7. **Rain, Rain, Go Away** 78
You stay dryer if you run rather than walk through rain.

8. **Natural Gas** 86
Six packs of Pop Rocks and six cans of soda is a lethal combination of gases that will make your stomach explode.

9. **Moving Pictures** 94
Will a decorative tattoo explode when undergoing a medical scan in a Magnetic Resonance Imager (MRI).

10. **Is Air a Con?** 102
It's more fuel efficient to run the air conditioner than to have the windows open.

11. **It's for you!** 108
A lighting strike on your house can travel down your home's wiring and kill you when you're on the phone or in the shower.

12. **Short Back and Sides** 116
A domestic fan can cut off your head.

13. **Hold on Tight** 124
A child can be floated into the air by holding onto a big bunch of balloons

14. **Hair-razing?** 130
If sparked, hair cream in an oxygen-rich environment will explode.

15. **Don't try this at Home** 138
If you swing hard enough, you can do a complete 360 degree rotation on a swing set.

16. **Mystery in a Can** 146
Everyday cola can clean off rust, dissolve a steak and a tooth, and clean the contacts on your car battery.

17. **What's Cookin'?** 154
Microwave power can cook your insides in a tanning unit.

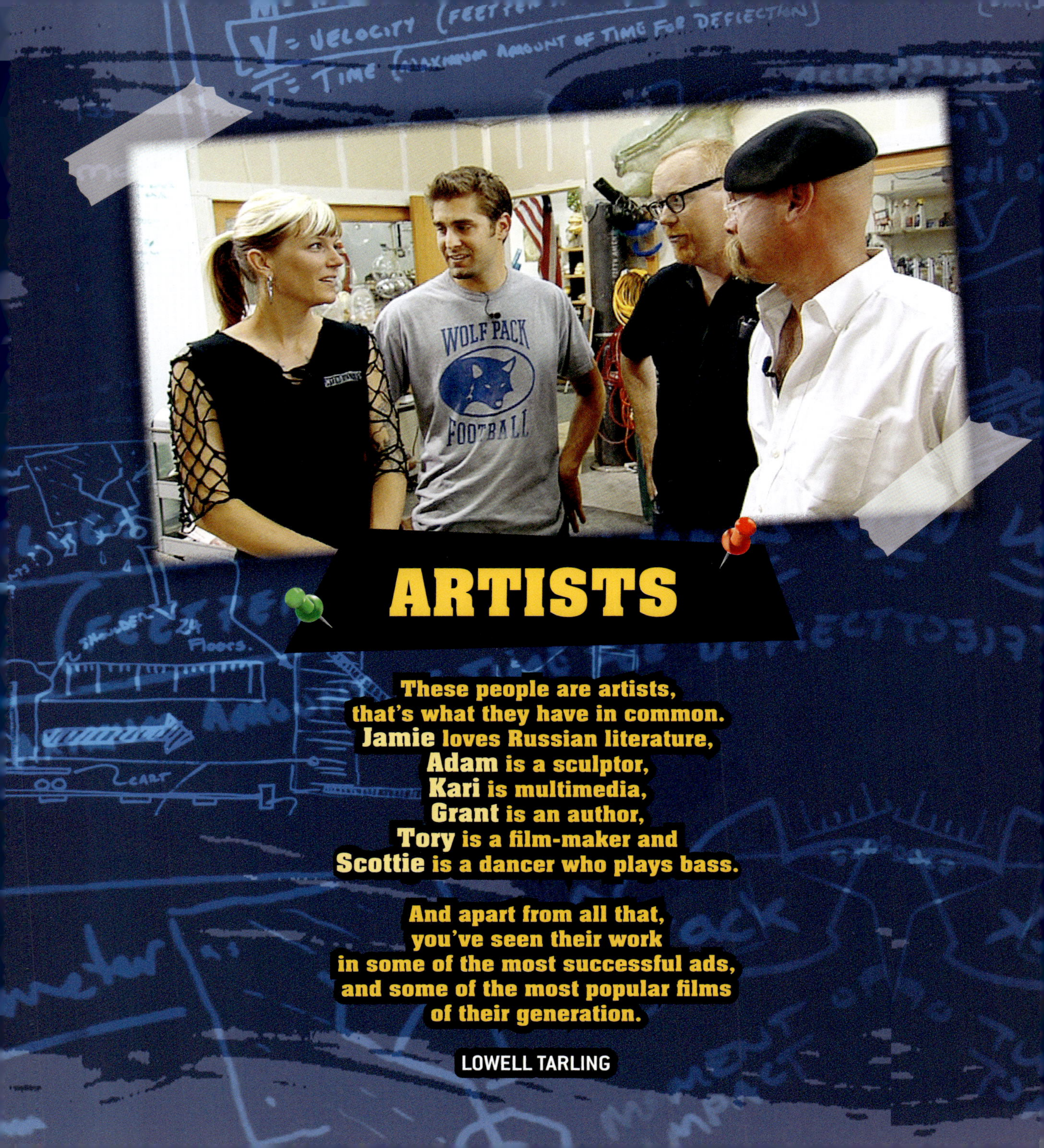

ARTISTS

These people are artists,
that's what they have in common.
Jamie loves Russian literature,
Adam is a sculptor,
Kari is multimedia,
Grant is an author,
Tory is a film-maker and
Scottie is a dancer who plays bass.

And apart from all that,
you've seen their work
in some of the most successful ads,
and some of the most popular films
of their generation.

LOWELL TARLING

Who are we?

A brief intro

LADIES AND GENTLEMEN ... THE MYTHBUSTERS!

FROM THE AUTHOR: **LOWELL TARLING**

The aim of this book is to re-create 17 Mythbuster myths for readers to enjoy in book format. My selection shows a bias for myths that might happen to you. For example, you probably boil water in your microwave oven, but you are unlikely to escape from Alcatraz Prison. So microwave oven myth in, Alcatraz out.

Because a book has the authority of being in print, its format is quite different to a television show. A book allows you to instantly check whether or not your tattoo could explode under x-ray, whereas the medium of television is more fleeting. I have quite a lot to say about writing from television, but I won't say it here. After all, this book is primarily about Jamie Hyneman and Adam Savage. It is also about the strong support team they have built around them: Kari Byron, Tory Bellici, Grant Imahara and also Scottie Chapman.

"I pull people from the local talent pool on an as-needed basis," says Jamie.

"Often I get them by referrals. I look for experience with a range of mediums, but otherwise I'm big on intelligence and work ethic. Putting together a crew is kinda like making soup: it's the combination of things that makes things work."

And if we can learn anything about inter-personal relationships from the success of the program, it is Jamie's understanding of how teams work.

"I pull people from the local talent pool on an as-needed basis." - Jamie

Maybe he'll write a book about it some day: Building & Motivating A Winning Team by Jamie Hyneman. I'd read it.

In Mythbusters – and especially on dangerous jobs – there is always a spirit of delegation and sharing the load. Right from the start, when Jamie was chosen to host the program, he told the producers that he was 'monosyllabic' and needed a co-host to make the show work. Jamie chose Adam, and where would the program be today without him? Together, they make the show. But what the viewer doesn't see is what happens off-camera, which is when the plans of attack are being formulated, and where the real leadership lies.

Mythbusters is actually a team of 20-50 people, many Australian. At the executive level there are people like John Luscombe of Beyond and Sean Gallagher of Discovery, without whom the program would never have existed. There's producer Peter Rees, who concocted the idea.

One of the greatest guitarists in the world, ethno-musicologist Bob Brozman, wrote the original music. Robert Lee narrates the show. President of the California Folklore Society and author of Ironic Bodies and Tattooed Jews, Heather Joseph-Witham is the folklorist.

These are all interesting people, but let me tell you about six you most often see on screen:

JAMIE HYNEMAN

Jamie Hyneman is a visual effects expert and founder of M5 Industries, a special effects workshop where Mythbusters is filmed. He is also known in the BattleBots circles for his robot entry, Blendo, who was tough.

A distinctive part of Jamie's appearance is his vicious moustache, dark beret and white lab shirt. He is known for his calm, logical, no-nonsense demeanor. Jamie has a degree in Russian language and literature.

He has had a variety of careers, including scuba diver, wilderness survival expert, boat captain, linguist, pet shop owner, animal wrangler, machinist, concrete inspector and chef.

Notable achievements include the Can Spitting Soda Machine from the 7-Up commercials, and the two-wheeled football shoe from NikeLab commercials. Jamie has worked as special effects technician for following movies: The Matrix Reloaded, Star Wars – Attack of the Clones, Star Wars – The Phantom Menace, Naked Lunch, Arachnophobia, Rude Awakening and Top Gun.

ADAM SAVAGE

Adam Savage has built: spaceships, Buddhas, puppets, rifles, sculptures and toys. He has worked in the Special Effects industry, and also as an animator, graphic designer, rigger, stage and interior designer, carpenter, scenic painter, welder, actor, writer, and now – television host. He's worked in metal, glass, plastics, neon, injection molding, vacuforming, pneumatics, hydraulics, electronics, casting, moulding, welding, brazing, machining, lathing, wood, animatronics and robotics.

Jamie's distinctive beret and moustache combo ...

Adam's handiwork has been seen on over a hundred TV commercials, and a dozen or so feature films, including Star Wars Episodes I and II, Space Cowboys, Galaxy Quest, Terminator 3 and The Matrix sequels. He has done research and development for toy companies, acted in commercials and films, and done props and sets for Coca-Cola, Dow Corning, Hershey's, Lexus, and a host of New York and San Francisco theatre companies. Adam is also a sculptor, of mixed media assemblage, whose work has been represented in more than 40 shows in San Francisco and New York.

Adam Savage has built: spaceships, Buddhas, puppets, rifles, sculptures and toys.

KARI BYRON

Kari Byron studied film and sculpture at San Francisco State University where she wrote, directed and starred in B-level schlock-horror films, sculpted intricate model dragons for Dungeons & Dragons fans, and graduated as an artist, working in sculpture and painting. Kari has had successful exhibitions at some of San Francisco's leading galleries.

Her sculpting skills led her into the world of model-making and toy-prototyping, and a job with Jamie at M5 Industries. It was here that Kari got her break with the Mythbusters team.

Kari also has worked in a kosher bagel store, and has posed as a store mannequin to foil shoplifters. Her interests include fencing and knife throwing.

GRANT IMAHARA

Electronics and radio-control specialist Grant Imahara is a former animatronics engineer and model maker for George Lucas' Industrial Light & Magic, where he worked on the movies: The Lost World: Jurassic Park, Star Wars – The Phantom Menace, Terminator 3, A.I. – Artificial Intelligence and The Matrix Reloaded.

In addition to operating R2-D2 (one of only a handful of official operators), Grant developed a custom circuit to cycle the Energiser Bunny's arm beats and ears at a constant rate. Grant is responsible for all the electronics installation and radio programming on the current generation of bunnies, and continues to serve as the bunny's driver and crew supervisor on numerous commercials.

He has a Bachelor of Science degree in Electrical Engineering and is the author of the book, Kickin' Bot: An Illustrated Guide to Building Combat Robots. His own machine, Deadblow, is a BattleBots champion.

TORY BELLICI

Tory first walked into Jamie's workshop in 1994 as a young film student from San Francisco State University looking to start a career in special effects. Jamie quickly put him to work sweeping the shop floor and running errands. But Tory quickly learned all he could and moved up the ranks. Three years later, he landed a job with George Lucas' special effects division, Industrial Light and Magic, where one of his biggest

projects was building models for the Star Wars trilogy. The Federation Battleships and Podracers in The Phantom Menace and Attack of the Clones are examples of Tory's work.

Tory's other special-effects work can be seen in the Matrix trilogy, Van Helsing, Peter Pan, Starship Troopers, Galaxy Quest and Bicentennial Man. One of his short films has appeared in the Slamdance Film Festival and on the Sci-Fi Channel.

SCOTTIE CHAPMAN

A former horse trainer, video-game tester and graveyard-plot telemarketer, at one stage Scottie Chapman and her special effects make-up and metal working qualifications saw her as the only legitimately qualified person in the show! Trained in welding, machining and metal fabrication she has built bridges, racing cars, amusement park rides and seaplanes. In the course of her career Scottie has worked in set-building, carpentry, model-making and architectural restoration.

She has taught Metal Fabrication in high school. She can also belly dance and fire dance. Her building talents have been used by San Francisco's famous industrial performance art collective, Survival Research Laboratories.

Scottie plays bass guitar in several underground punk/noise bands including the Wizards of Ozzie, a Black Sabbath cover band.

M = MASS (SLUGS
MAIN VARIABLE
V = VELOCITY (FEET PER SECOND)
T = TIME (MAXIMUM AMOUNT OF TIME FOR DEFLECTION)
man
Newton/sq meter
crack
MOMENT
IMPACT

The Pioneer

Aussie Made

TWISTED SCIENTIFIC INNOVATION

Most people don't know that the award-winning Mythbusters program is an Australian TV show. But check the company business card and you'll note it has offices in London and Dublin, as well as Artarmon in downtown Sydney. It therefore comes as no surprise that the entire Mythbusters concept was conceived in the Artarmon Beyond studio, where the post-production work is done to this day.

The head of production is John Luscombe, who was involved right from the start. It was Luscombe's belief in the concept that lead him to successfully pitch the show to Discovery Channel, from which it has gained a regular place on mainstream TV.

The idea itself stemmed from director, producer and scientific consultant, Peter Rees (*pictured opposite*). Says Peter: "I'd been making science documentaries for Discovery Channel in Australia for 10 years, and in 2000 I was looking at all kinds of ideas and I figured that everyone who'd previously done shows about urban legends – and of course there's been many – have re-enacted the myth and got experts to talk about any significance they have gleaned from folklore and history. I suddenly thought 'What if they put these stories to the scientific test?'

"We don't just re-tell the legends, we put them to the test." – Peter Rees

"And so the pitch for the show was: 'We don't just re-tell the legends, we put them to the test.' That one line got the whole show up. You may have heard the story before, but have you seen anyone test it? We had a bit of trouble deciding on a title. We had all kinds of names (like Tall Tales & True, etc) then one day a bunch of us from Beyond were sitting in a meeting and the most obvious thing to call it was Mythbusters. John flew that, and Mythbusters it was. So Beyond took it to Discovery. They liked it and they green-lit it for three episodes. Next we went on a talent search, which took a period of six months to find the right presenters."

IRREVERENT AUSTRALIAN APPROACH

Despite being an Australian-driven production, Mythbusters is filmed in San Francisco. "During the first few years the entire crew was entirely Australian, which was fantastic," says Rees, "just like a party. I think we made a bit of an impact."

Of course, the thought arises that, apart from parochialism, who really cares about the program's country of origin? It is clearly a wonderful collaboration between all parties, but Peter believes that the content is largely explained by the Aussie input.

"It wouldn't be the same show if it were made by Americans," he says. "One of the main reasons the show works is because it's got an irreverent Australian approach to everything, and while the Americans initially found that incredibly frustrating (and still do) it has really set the tone for the entire show. It has an irreverent view of products and authority, and a can-do approach. I think this has been really important in setting the tone. Americans are very deferential to authority, Australians aren't.

So when we did a story suggesting that you can get around drink-driving, it would have been totally humourless if Americans had done it because they're apologetic about

"Rees is the third Mythbuster, or maybe even the first, since the show was his original creation and carries his indelible stamp of twisted scientific innovation." – Keith and Kent Zimmerman, Mythbusters – The Explosive Truth Behind 30 of the Most Perplexing Urban Legends of all Time.

breaking the law. Australians were all putting in the boot! Our American presenters, Jamie Hyneman and Adam Savage, often say, 'This isn't going to work' but the Aussies consistently reply, 'We don't care, we just have to try it!'

"We engage the audience on a level which they can go out and do. They can go out in the traffic and test their air-con and their results will be just as valid as ours. And although most television broadcasters think the appeal of any show is star-power, when people get on our website they're talking about their personal experiences, which is what's important to them. That's where the appeal of Mythbusters lies.

"When the show came out in Australia everyone was saying, 'There's this great American show!', while John Luscombe (from Discovery) and I were going, 'You should publicise this as an internationally successful Australian show!' All crew members apart from the hosts are Aussie! The only thing the Americans found tough is that we obviously have a very different sense of humour. Adam once turned to me and said, 'You know, I've suddenly realised – abuse is a form of affection in Australia!'"

Tory's love of special effects made him a perfect fit ...

IDEAS

Some ideas for the program come from the Discovery fan-site, and Adam and Peter are both avid readers. Says Peter: "Our main sources are the things that are happening in contemporary news, and we try to reproduce those circumstances. If that doesn't work we try to reproduce the results.

"Everyone wants to read the books that Jan Harold Brunvand puts out about urban legends – he writes these things up in a very scholarly way – but I think we've probably created a few of our own."

(Brunvand is a Professor Emeritus of English at the University of Utah in the United States and is best known for spreading the concept of the urban legend. Before his work, folk tales were associated with ancient times or rural cultures which Brunvand applied

to stories circulating in the modern world. He is the author of several well-known books on the topic of modern myths.)

Peter continues: "As a collection point for raw weird stories, we are unsurpassed. We have our own fansite www.mythbustersfanclub.com and if some folklorist is not writing a PhD on that fansite they should be. Even though we're not the highest-rated show on the Discovery Channel, we normally have up to 10 times the web-traffic of any other show simply because people are fascinated."

Peter explains how the shows are conceived.

"I'll come up with the theoretical framework and say, 'Jamie, we have to build a hybrid rocket system – what do we need to do? Is this possible? Is this feasible?'. And that's where he and Adam come in. Adam and Jamie's brilliance lies in the fact that Jamie, for instance, can go down to the local plumbing shop and see a rocket: he can 'see' all parts he needs and he can make it within three days and it will work."

"The one I really want to do is the myth that when people die their body supposedly loses 21 grams in weight which some people say is the soul departing the body." – Peter Rees.

PRESENTERS

"Jamie and Adam's contribution to the show is invaluable because they are brilliant builders," says Peter. "Once you set them to it there's no holding back, because Adam and Jamie can obviously build anything with anything."

Mythbusters took on a life of its own. When Luscombe and Rees were trying to figure out how to present the show, the original idea was to first have an investigator who would interview the people involved in the original myth. (Where that would not be possible, they planned to use on-screen interview-style evidence.)

The second presenter would then test those experiments in a workshop, an approach which is reflected in the first three episodes. They wanted someone who would be able to build anything and have a workshop to build it in – one that was big enough to double as a television studio. The final criteria was that he / she would be a good television host.

They conducted a search for suitable hosts but no special effects crew was big enough to support the project, so the Beyond team looked overseas, where Jamie and Adam had been in the running from the start because of their respective backgrounds in special effects. These were early days, indeed.

"Mythbusters was created by Peter Rees," Adam recalls. "Peter produced the show Beyond 2000 out of Australia and he had interviewed Jamie and I about a robot we had in the original Robot Wars back in the mid-'90s. Apparently a good producer never throws away a phone number, because in the Spring of 2000 he called up Jamie and asked him if he had an interest in hosting this show he was trying to cast. Jamie called me, we sent in a demo reel, and apparently they loved it. 'These were just the geeks we were looking for' was what we heard back."

Jamie confirms Adam's recollection (a rare event): "It was the idea of our producer Peter Rees. He had interviewed me some years ago during Robot Wars when I had a robot called Blendo, which was instantly killing all the other robots. I was therefore somewhat notorious, so Peter spent a little time with me and when he had the idea to do the show he contacted me. I thought I could do the show but not carry it by myself because I am not all that animated. I called Adam, who was an ex-employee of mine and who was the liveliest FX guy I knew. We did a demo tape, and the rest is history."

"Jamie and Adam were selected for their building abilities," explains Peter, who had met Adam 10 years previously when Peter worked with Jamie on Robot Wars story. "Jamie was the first person we approached to be our presenter. When I asked him he said he didn't think he'd be very good on television because he reckoned he was a bit monosyllabic, but he said 'Ah've got … a friend … who … like … he does most of the … talking'. And that friend was Adam."

Both boys are hoarders, ensuring everything is available for any test … you've just got to find it.

In the summer of 2002, Jamie and Adam spent five weeks with the Aussie team shooting nine episodes (or three-and-a-bit one-hour shows) for the Discovery Channel. There was only five on the crew and they did everything themselves. Peter designed the experiments and, when Jamie and Adam executed them, everyone on the set could see that these guys clearly had a lot of mechanical savvy between them.

At this time, the Beyond crew still hadn't got over the idea of having an investigator and a single builder, but after the first part of the shoot they were so impressed that Peter argued for a change in format. "This is the entire show right here," he said, "we don't have to add a single thing."

Grant's versatility has been a huge asset as the team's experiments get grander in scale.

Everyone agreed, and after clinching Jamie Hyneman and Adam Savage as presenters, the new focus became the novel techniques validating weird and wonderful modern misconceptions.

As the Beyond team watched Adam and Jamie prove, disprove and / or replicate the results of the myth at hand, it was obvious to all that their creative problem-solving was consistently impressive. As a bonus, even their bickering was interesting. Their disagreements over methods was intriguing to a point where Adam and Jamie's respective personalities blended into 'good television', and the first three episodes went to air in 2002 with mixed results.

But Mythbusters have since become an international sensation. They have expanded the team to include actor-artist Kari Byron, electronics and radio-control specialist Grant Imahara, model builder Tory Bellici, metalworker Scottie Chapman and artist-globetrekker Christine Chamberlain.

"I'd like bring in more people who have specific skills whether it be in rocketry or chemistry or whatever." – Peter Rees.

WHEN IT DOESN'T WORK?

The great thing about the show is that if it's a dismal failure (so long as it's a spectacular dismal failure) the Mythbusters still have a show. Out of everything they have attempted, there are probably only five stories they couldn't complete.

The worst was the jet taxi story – they were trying to flip the car in the slipstream of a 757 jet. Jamie was standing on the runway with his finger on the remote control of a full-sized taxi; meanwhile, the producers were making calls to the Sydney office trying to work out if they'd been granted insurance coverage, which they didn't get. So they made a story out of not doing it.

Another one they didn't get off the ground was the 'Mind The Gap' myth at train stations, where you're not supposed to stand between the yellow line and the edge of the platform. It is believed that when the train comes past it could supposedly create enough suction to pull you onto the tracks. The Mythbusters still want to test that but can't get any train company, anywhere in the world, to let them put their crash test dummy between the yellow line and the passing train. *(They thought they might try that in Australia, because producer Dan Tapster spotted the myth spelled out on a sign on the platform at the Penrith Railway Station on the outskirts of Sydney. Maybe the Australian Road Traffic Authority (RTA) will let them do it. And maybe not.)*

MYTHBUSTERS: THE MOVIE?

Peter would love to do Mythbusters: The Movie. "I'd like to do Mythbusters IMAX where we take everything full-scale," he says. "I'd also like to try some things that we would never be allowed to put on Discovery Channel, like 'Do black men really have bigger penises?' There's a whole bunch of stories that there's no way Discovery Channel is going to let us do.

"We've had semen onscreen a few times. Son of a Gun is a US Civil War story in which a bullet allegedly grazed a soldier in the scrotum, enough to pick up his sperm before burying itself in a woman's womb. And so she fell pregnant – like a virgin birth. Because this experiment required a large sample, to test this we acquired semen from the entire crew, so on the morning of the shoot every male crew-member had to bringing in his little baggy with live swimming samples, after which we had to attempt to impregnate a ballistics gel womb by firing a bullet."

Scottie is the only Mythbuster with actual qualifications in anything.

"Apparently a good producer never throws away a phone number, because in the Spring of 2000 Peter called up Jamie and asked him if he had an interest in hosting this show he was trying to cast. Jamie called me, we sent in a demo reel, and apparently they loved it. 'These were just the geeks we were looking for' was what we heard back." – Adam.

WHAT'S NEXT?

Although the Mythbusters don't do sex, they may move the program into more controversial areas in the future … political and religious themes, for example. One suggestion is a re-enactment of US Vice-President Dick Cheney's description of the circumstances which caused him to accidentally shoot his companion, a prominent Texas attorney. Could it have happened according to his story?

Peter Rees would like to re-enact Cheney's testimony and produce a type of program which would position Mythbusters in a whole new category of political television journalism.

The religion angle would be equally as controversial. Peter suggests a religious myth they'd like to put to the test would be titled 'The Power of Prayer' – can prayer influence an outcome?

One published study says yes, but the Mythbusters would like to find out for themselves. They propose that a random number generator could be used to test whether prayers can influence the outcome. Tory has agreed to pursue this theme.

Boys and their toys …

Even more significant is the notion of the division between the spiritual and physical body, and Peter would like to test a study (by an American religious minister in 1906) that the soul leaves the body at point of death. He says, "I really want to do the myth that when people die their body supposedly loses 21 grams in weight – which some people say is the soul departing the body."

It's onward and upwards for Mythbusters. As an example that the sky is the limit, they'd like to test the Moon Walk myth – did it happen or not? Peter says, "I wouldn't mind trying that, but the only way we can do it is if someone sends us to the Moon – and someday someone might just do that, you never know!"

PETER REES ON JAMIE HYNEMAN

Jamie and Adam are totally self-taught, but Jamie has had a particularly diverse background – he owned a pet store for a while and had a pet lion. Jamie is driven by building weird things. He grew up on an apple farm in Indiana, where his father still lives. Then he ran a diet business. He has a degree in Russian Literature. Jamie has done all kinds of stuff, he had a patent on a nano-muscle that was going to be put into action figures so G.I. Joe could actually move. He's into hybrid cars and he wants to build himself a mechanical exo-skeleton that can lift heavy weights. He's very proud of himself physically and he's pushing 50 this year. Jamie has total control over every aspect of his mind and body. You could run a sword through Jamie and he wouldn't flinch if he didn't want to flinch.

PETER REES ON ADAM SAVAGE

Adam is the total opposite, which is why the combination works. Adam has no formal education beyond high school but he's intrinsically interested in anything and everything. Adam's background is making theatre props and subsequently props for feature films. He worked on one of the Star Wars films and is an expert model-maker who works incredibly fast. He's all boom or bust, whereas Jamie's much more prepared. Jamie says Adam is like a toy – when the batteries are in and the switch is on he's absolutely unstoppable, but every now and then the switch turns off and Adam is fast asleep in the corner. Adam always has the latest in technology. He collects things. We have constant trouble because he wants to keep every single item from every story we've ever done. Most people might think it's with a view to selling it down the track, but no … Adam has at least three storage spaces full of boxes of crap. He's just bought a speaking monkey head. One day he said to me, 'I just bought a pair of stilts off the web'. Why would he want stilts?

SASCO
Upright

Sink or swim, and on national – and international – television no less. The team's commitment to putting myths to the test, no matter the scale, has made them a worldwide phenomenon.

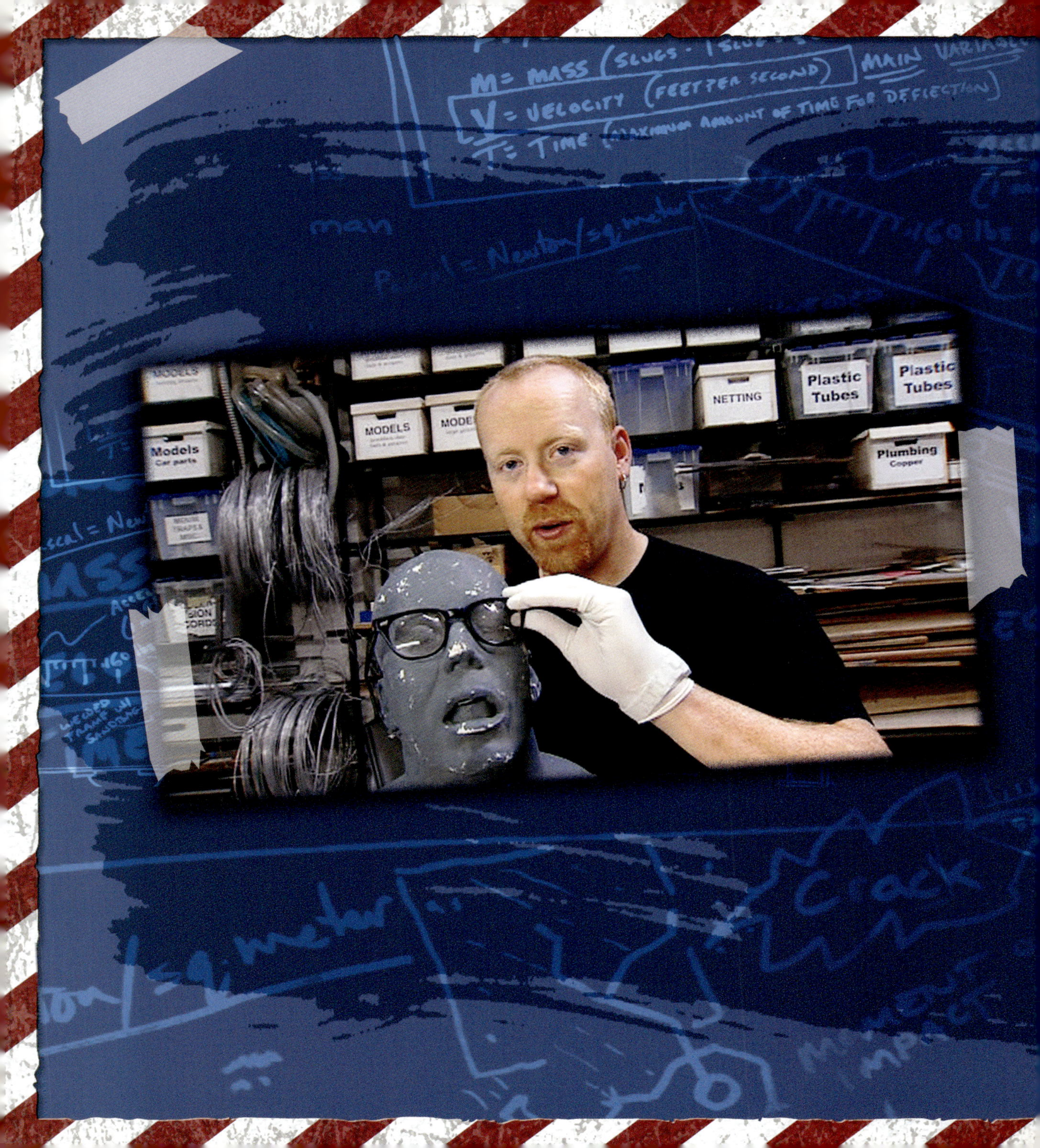
M = MASS (SLUGS
V = VELOCITY (FEET PER SECOND)
MAIN VARIABLE
T = TIME (MAXIMUM AMOUNT OF TIME FOR DEFLECTION)
man
Pascal = Newton/sq.meter
MODELS
Models
Car parts
NETTING
Plastic
Tubes
Plastic
Tubes
Plumbing
Copper
Crack
sq.meter

Light your Tongue

Chapter 1

SEARCH FOR A STUD

The Myth: ***A man was hit by a bolt of lightning. Was the lightning attracted to his metal tongue stud?***

The idea for the Tongue-Stud Lightning Test came from a news story about a British backpacker who claimed she had been struck by lightning while travelling in Thailand. After reading the story, producer, Peter Rees says, "How lightning works is not a completely understood phenomenon, so it's interesting from a scientific perspective. You've got the whole science of lightning here – Mythbusters is all about the science."

In America alone, lightning kills about 100 people per year. In August, 2003, Matt Thompson was hit by lightning in the hills outside Keystone Colorado. He almost died. It was big news on the internet after it happened, with chatrooms debating the question: was the bolt of lightning attracted to his metal tongue stud?

This theory reasons that the bolt is attracted directly to the stud in the same way as it is drawn to a lightning rod. "It sounds kinda silly," says Jamie, shaking his head.

"Lightning is really not predictable." – Jamie

When being loaded into an ambulance with a fried tongue, you can bet Matt Thompson didn't think it was a "silly" myth. Still, Adam and Jamie are sceptical.

Two human heads – modelled on Adam's, actually – made from ballistics gel, will be mounted near a generator capable of blasting out lightning at one million volts. Only one of the heads will be pierced with metal studs. The other will serve as the control. To find out whether the metal attracts the charge, Adam and Jamie will fire lighting bolts at Adam's ballistic gel heads and observe whether it is drawn to the stud. To achieve this, Adam and Jamie have enlisted the help of Delta Star West Inc, manufacturer of medium-power transformers, mobile substations and industrial grade electrical machinery.

- **The rig.** The first job is to create a small rig on which to affix Adam's heads. Something metal and round would be ideal – a wok? Adam and Jamie visit a kitchenware store where Jamie buys 10. He likes them because, unlike a spike, the round surface won't gather the electricity around a single point.

- **The heads.** Meanwhile, Adam has taken a casting of his own head (open-mouthed and startled!) which he is filling with rigid foam. He will cut this prototype in half from which he will make 10 ballistic gel heads. Sawing the image of his own likeness in half appeals to Adam: "There's something strangely satisfying about doing that!"
- **Fixing heads to the rig.** The two gel-heads are now the focal points. To secure each dummy head, Jamie attaches non-conductive spikes to the top of each upside down wok. These are made of plexiglass so as to not attract electricity in their own right. A metal stud – the "barbell" – is fitted to one dummy's lower lip.

They are using Delta Star West's gigantic bank of capacitors to fire electricity across a wire, through an electrode at the gel-heads which are grounded to complete the circuit. The tongue-stud lightning test begins. Adam and Jamie are about to create mini-lightning at a flick of a switch.

A million volts, stored in a capacitor, will discharge down the wire, jumping to the ground through one of the two heads.

ADAM COPS A TONGUE STUD

"This was one of those stories where we said, 'To really test this out properly someone's going to have to get a tongue stud inserted'. And Adam tends to volunteer for those things because Adam doesn't mind any kind of humiliation," says Peter Rees.

Adam went all the way, here's how:

In the spirit of this myth, and for no reason connected with the experiment, Adam volunteers to get his own tongue pierced. To do so he calls on Body Manipulations, San Francisco, which prides itself on being America's oldest piercing studio (1989).

He is attended by Joann Wyman, a woman with arm-tatts who is extremely composed. Jamie accompanies Adam whose only involvement is to make jokes. Says Joann, to a perplexed Adam: "Let's see your tongue..."

He sticks it out, she says, "Yes, we can put some jewellery in there for you."

Adam is clearly nervous: Have you heard any myths about tongue piercings?

Joann: I've heard some realities about tongue piercings, I haven't heard any myths. What do you have in mind?

Adam: There's a story going around about a woman in Britain who was apparently struck by lightning and she said it was because she had a tongue piercing.

Joann: None of my clients have ever been struck by lightning. She shows him a range of beads, and offers some gemstones "... if you want to get something really flash".

Adam: What do you think Jamie, should I go flash or plain?

Jamie: I think you should go for maximum conductivity.

Joann: Sit your glasses right there. Are you ready? It's very quick, very easy. Swallow one last time, breathe easy. Now give me your tongue. I won't hurt you. I'm not going to pull on you, I'm going to make sure I've got you right where I need you. I'm just getting you lined up. Piercing is real quick. Here you go – click - and out.

Done. The jewellery is in. She takes the clamp out and screws the bead on the bottom.

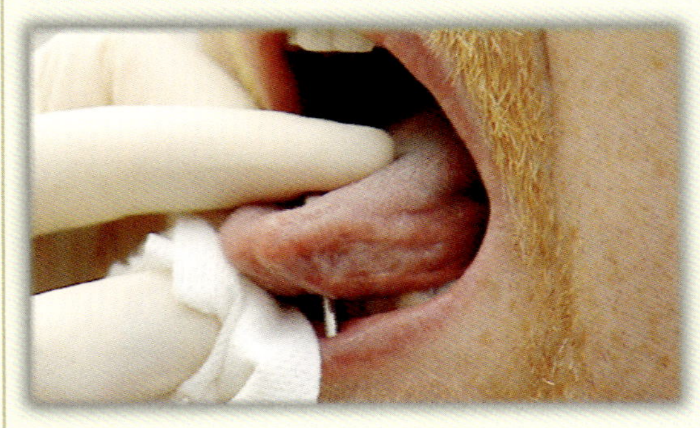

VERDICT

Adam: "That was not tho bad, oh – but now I'm talking all funny. I feel as if I've got a barbell stuck in my tongue man. Thee thells thea theels by the thee thaw!"

THE EXPERIMENT

The Mythbusters press the switch:

CRAAACK!

Result: The bolt fires into the head without the stud. A puzzled Jamie summarises. ***"We learned a couple of really important things from the first test:***

(1) The lightning bolt went to the top of the head and nowhere else on the electrode, which tells us that our rig is nicely set up; it's pretty neutral and we are getting current to go through the head itself and not through the seal.

(2) The second most important thing was the discharge actually went to the headwithout the piercing, so the myth's not looking too good right now. But the testing has just started.

In the interests of accuracy, Adam and Jamie swap the heads on the rig to ensure that the lightning doesn't favour one direction over the other. Everybody heads for cover when they press the switch a second time. **CRAAACK!** This time it does hit the stud head. They do it a third time – **CRAAACK!** – resulting in a total of two strikes on the pierced head, one on the control head. Will swapping them back make a difference? They try – everybody heads for cover once more and it's the pierced head that gets the bullet again. But although the charge often goes to the stud head, it never touches the metal. The Mythbusters won't be satisfied until they can see the lightning hit the stud directly. Jamie wants to up the ante: "To see if we can get something to react, all we've got left to do is to put a whole bunch of tongue studs in the stud-head, some really big ones!"

Jamie's gel-head is now subjected to more studs than a pick-up bar. Will all these extra studs influence the lightning? The tests answer the question with a "No". To keep things lively, Adam and Jamie add the mother of all studs: a metal door knob. A direct hit at last. It knocked the doorknob right out of the face!

RESULT:

LIGHT YOUR TONGUE

CONCLUSION:

It is dangerous to walk through a lightning storm with really big metal bolts or a doorknob in your mouth.

- The Mythbusters are unconvinced that jewellery studs attract lightning.
- While the zap seemed to prefer the stud-head, a lack of consistent direct hits suggests some other unknown factor was involved in the outcome.

Jamie: *"Lightning is really not predictable. We could have done this 1000 times and still not gotten that same result."*

Adam: *"That may be the case Jamie, but we do know metal can have an effect depending on how much there is and where it is on the body. And we did get a strike right in the doorknob on this one."*

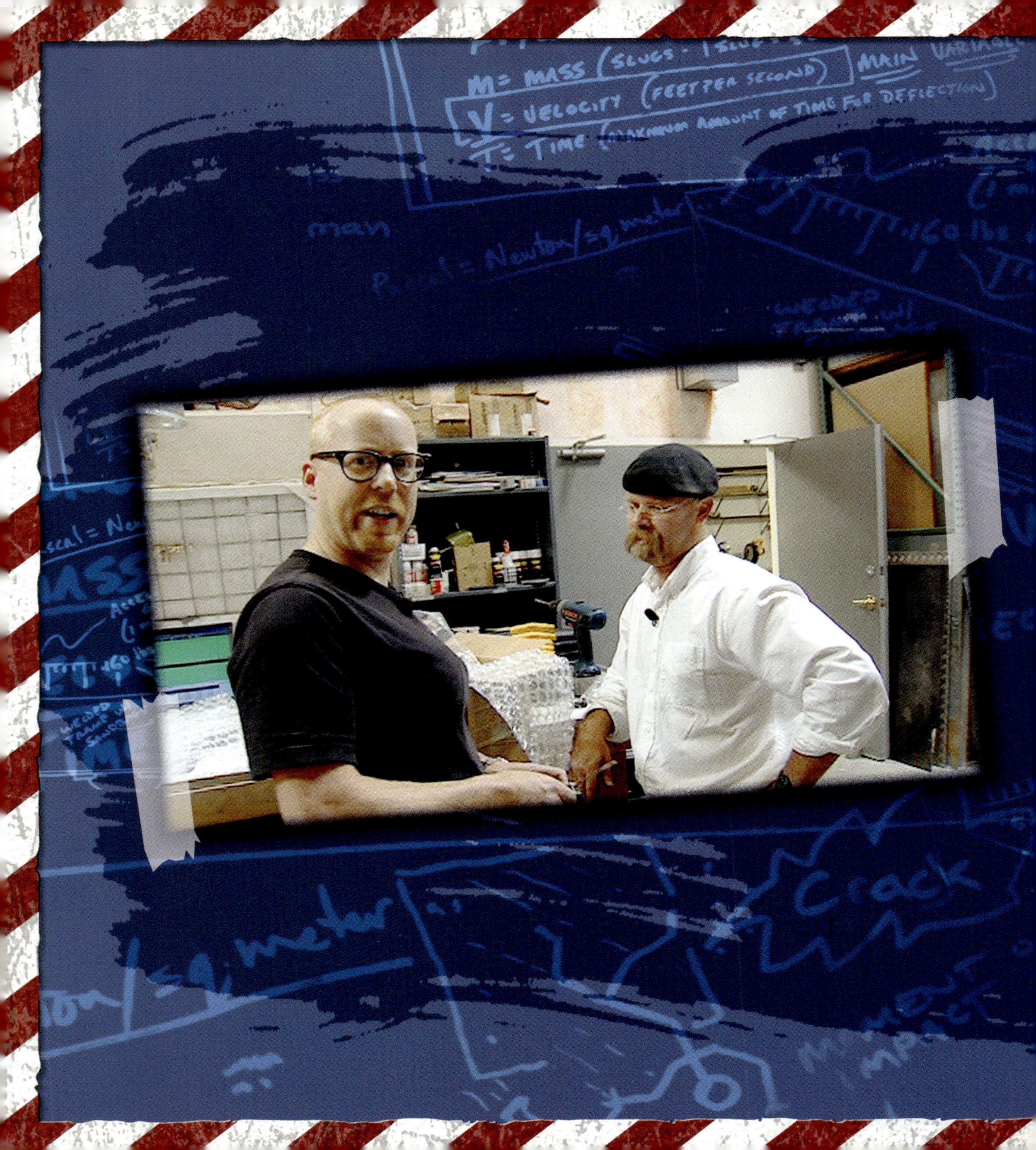

M = MASS
V = VELOCITY (FEET PER SECOND)
T = TIME
MAIN
man
Pascal = Newton/sq. meter
160 lbs
Crack

Is this a Bust?

Chapter 2

EXPLODING BREAST IMPLANTS

The Myth: ***When you go up in an airplane the reduction in pressure will explode your implants.***

Adam walks into the Mythbusters studio carrying a box which he immediately opens exclaiming, "Jamie, I've got breasts!" A startled Jamie looks up.

"Check these out," says Adam, playfully holding two white silicone-filled shells against his chest. Each bears the words, "Sample, not for implant."

"I take it one size fits all," says Jamie ruefully. "It's not like 'bigger' or 'smaller'."

Peter Rees explains what's going on, which is the fear that breast implants may burst under pressure. "The breast implants story is about air pressure and different styles of breast augmentation. We went back to the inflatable brassieres of the 1950s (the soft pillowy foam sometimes put in bras to bulk them up). We took all those things onboard and also included the silicon breast implants."

"I've got breasts! They make me feel like a woman." – Adam

But here's the rub. Adam not only gleefully wears the implants, but he keeps feeling himself up throughout the episode which, even edited, is a somewhat disconcerting sight.

"A little known side-story on that shoot was that we had difficulty getting this episode through *Discovery* because Adam was seen palpitating his breasts on many of the shots," Peter recalls. "And there's a really good scientific reason for that, which is that even with a slight change in altitude, and even though they wouldn't burst, he should feel tightness as the external air pressure decreases. So he's fiddling non-stop with his implants while wearing a leopard-skin bra, and *Discovery* being *Discovery*, has a bit of a problem with someone constantly touching their breasts – silicon or no silicon!

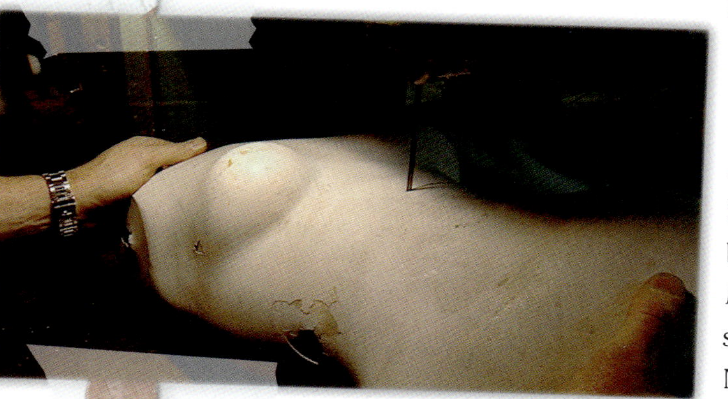

Breaking up is so hard to do ... Adam gets to work on our dummy.

"There's an old adage that 'you can show a breast on Discovery Channel if it's got a gun-shot wound - but under no other circumstances'. So Adam fiddling with his breasts raised alarm bells and large sequences of that story got cut. *Discovery* didn't understand that there was a scientific justification for Adam's discomfort." No-one else believed him, either.

While almost 100,000 patients a year have breast augmentation for cosmetic purposes, some are worried.

Folklorist Heather Joseph-Witham was part of the initial Mythbusters team.

"The basic story is that a woman with silicon implants was on an aeroplane which had a depressurised cabin because it was a local flight," she explains. "As the plane went higher, the implants got bigger and bigger and ***bigger***. As she was walking down the aisle to the lavatory to see what was going on, they just exploded."

Plastic surgeon Dr Gregory Geogiade, from the Duke Medical Centre, has performed thousands of procedures over 25 years, but he's never taken this myth seriously. "I've had no first-hand experience of a patient with a 'blowout' at altitude, but lots of people are worried about it," he says. "However, I am frequently asked by patients who are going to skin-dive, what happens to implants in the deep?"

"So we've got implants – right?" says Adam, slipping them under his blue v-neck jumper and feeling himself up. Again. "I don't know what they are supposed to feel like, but these feel pretty good. They make me feel like a woman."

Jamie: "You're giving me the creeps. Take those off!"

Adam takes them off and juggles them, "What else are we going to need Jamie?"

To explode the implant myth the Mythbusters are going to need:

1. A chamber that will handle both positive and negative pressure as well as a vacuum
2. A pressure pump
3. To make it look good, the implants need to be positioned into a body of some sort.

- **The chamber.** To reproduce the altitude pressures that may cause an implant to expand or explode, Adam and Jamie are seeking a tank to convert into the chamber. Jamie's first job is to find a chamber capable of doing the job, so he calls on familiar territory - the scrap metal and high-tech surplus section of the local junkyard.

"This is like a Sci-Fi wrecker, I love this place," he says. "Is that the world's most expensive garbage can?" Adam laughs, pointing to an object locked between containers and beer kegs. But Jamie has found something which better suits his needs, he strokes it and says, "It's a beautiful piece of artwork, I get such a kick out of finding this stuff. Look at the thickness! It probably cost somebody maybe $30,000-$50,000 new but I'm sure it's going to be priced by the weight of the aluminium."

"God bless them!" says Adam, as the wrecker confirms Adam's deep understanding of scrap metal dealers with the words: "I don't care what it was, just tell me how much it weighs."

LET'S GO "ALL THE WAY"

Adam volunteers to test the inflatable bra at 2500 metres. Wearing a fake leopardskin model, Adam lies inside the pressurised chamber, facing the depressurisation slowly and in increments.

Adam reports in: "Ground to mission control, my breasts are definitely getting firmer, but nothing's exploded unfortunately. They're definitely firmer at 8000 feet but not by much."

"Do you want to go all the way?" asks Jamie.

"Not on camera," laughs Adam.

"Shall we take it up to 40,000 (feet, or 10,500 metres)?"

"Let's do that," Adam agrees, "But not with me in them."

Jamie is keen to push the inflatable inserts to the limit. There should be obvious changes at 10,500 metres.

Verdict: Even at an intolerable altitude the inflatable inserts are in no danger of exploding.

• **The torso.** Using a mannequin, Adam's first job is to make up a torso from ballistics gel and suspend the implants very close to the skin surface. Being clear, the gel will allow the Mythbusters to see through the 'skin' and observe any movement or expansion of the implants. Adam saws the mannequin in half, and a heat vacuum appliance is used to make a plastic mould of the torso. Then Adam places ice against the skin of the mould to make the outer layer of ballistic gel harden, but not the rest.

This results in a flesh-like barrier, like a layer of human skin, about 5mm thick. The skin layer takes about 10 minutes to form and then it's just a matter of putting the implants into place.

Ever-holding the breasts, Adam says, "I'm going to try and implant these fake breasts into a fake woman." Being lighter in density, he hopes the suction will keep the silicon implants in place. When he returns 20 minutes later, Adam is well satisfied with the result. Adam's gelatin torso has set. The implants are holding. "It's a thing of beauty!" he cries.

• **Welding.** The important job is to make the chambers as airtight as possible. Some of the chamber's portholes are welded shut, while others are replaced by Jamie. Finally Jamie fits an air pressure gauge which Adam calibrates with an easy-to-read altitude guide for the gauge, indicating 12,000 metres, 2500 metres and sea level. "The chamber is sealed," proclaims Jamie. "Are we ready?"

One man's trash is definitely another's treasure, especially if the Mythbusters can get their hands on it.

THE EXPERIMENT

As a parallel experiment, Jamie places an implant in a jug of water with measurements, which will be also subjected to pressure. The current level is noted and marked. Should the implant expand, it will be easy to see because the water level will rise.

2500 METRES: For the first test the implants are subjected to typical in-flight cabin pressure which is equivalent to around 2500 metres above sea level. ***Result:*** Jamie observes that the bubbles that were on the surface of the implant have gotten larger. This has raised the water level in the jug a miniscule amount, whereas the breast has not visibly expanded.

Despite being under extreme pressure, the implants simply did not explode under duress.

10,500 METRES: The big test comes as the chamber matches an altitude of more than 9100 metres, an impossible height for humans unless flying in a pressurised cabin. Anybody in there would probably be dead within 10 minutes. Their lungs would be covered with fluid, their head would compress, and other problems would occur … none of them pretty.
Result: Despite the negative pressure of high flying, the implants appear unchanged. Says Jamie, "We've seen the implants at about 8000 feet and we bumped it up to about 35,000 feet and in both cases we've seen expansion of little air bubbles to some degree in the water jug but the implants themselves have shown no significant increase in volume."

THE INFLATABLE BRASIER

In the 1950s there was a myth that a national sales manager for the inflatable bra was on a flight in a de-pressurised plan. She was wearing the product and as the plane got higher her bra got bigger and bigger and bigger. Seeing this, the passengers started to become disturbed, so she spent the rest of the flight in the cabin with the pilot to keep her out of sight.

Adam's enjoyment of this particular experiment extended as much to style as substance.

ADDITIONAL AIR:

Adam suggests that injecting the implants in the chamber with air might force a failure, so he needles 3ccs of air into one and 10ccs of air into the other. Jamie turns the pressure up to 10,500 metres. They also place an orange balloon within the same chamber.

Result: The balloon doubles in size but the air inside the implants does not appear to expand.

SEA PRESSURE:

Jamie hooks up the small chamber to emulate the high pressures of deep sea diving. He didn't figure on explosive results. A seal blows – *Booom!* No-one is hurt, but Jamie had a close call. So it's back to the large chamber for another low pressure altitude test. Within a few minutes the implant is subjected to the equivalent of a rapid trip from the ocean's floor to the summit of Everest.

Result: No noticeable difference.

CYCLES OF INTEREST

"We don't invent the stories, they just 'appear'. These stories go through cycles of interest as they're flying around the planet and around communities. And so they will be reinvented and re-told and unfortunately many news outlets run these stories without fact-checking them, before they get to air. And that's when we identify them and say, 'That's a classic for us'." – Peter Rees

RESULT:

IS THAT A BUST?

BY THE WAY ...

Scientists have been studying the rigours of pressure on the human body for many years. Dr Richard Vann, from the Centre for Hyperbaric Medicine, Duke University, California, knows all about pressure and the concern that people have with implants.

Dr Vann says: "We've had questions whether there are any hazards with breast implants with regard to diving, and as a result we decided to do a study, and we found that basically anything the patient could survive wasn't going to be an issue."

The results of Dr Vann's implant pressure test were convincing. Two implants that were initially the same size were expanded by about 50% to show what an expanded implant would look like.

Result: At 50% bigger, the implant could still double or triple its total volume before the shell was ever going to break. This is never going to be an issue. The patient would have been dead from nitrogen bubbles to the brain long before the implant got to this size. The myth of the exploding implant appears to be well and truly busted.

M= MASS (SLUGS
DEFLECTION)
MASS

A Clean Kill

KILLER APPLIANCES IN THE BATH

Chapter 3

The Myth: ***Water conducts electricity to the point where an electrical appliance dropped into a bath will electrocute, and probably kill, the person in it.***

"What's with the bathrobe?" asks Jamie.

Adjusting his blue robe, Adam replies, "This time we are tackling one of the classic Hollywood movie myths, which is simply: if you threw an electrical appliance into my bath, I'd get electrocuted. This grisly death is a favourite with film makers. For example, the horror movie *Chucky*."

"Well that means *appliances"'* says Jamie, which means a visit to a local second-hand shore. "Our favourite haunt – let's go!"

"Everyone wants to know whether we're just busting myths or actually confirming them – this one 100% confirmed." – Adam

The store carries a huge range of electrical goods. Television sets, digital clock radios, fans, sound equipment, sewing machines, calculators, computers, standard lights – the choice is virtually unlimited.

While chatting to the shop assistant, Adam and Jamie comment that most up-to-date models are fitted with a device known as a "ground-fault interrupter" (GFI) which is designed to give protection from an electrical shock by interrupting a household circuit when there is a difference in the currents in the wires. On detecting any sudden change in the flow of the electricity, it immediately shuts off the current. This means the newer appliances should immediately turn off, when they are dropped into a bath tub.

Don't press the issue ...

"Older appliances don't have ground fault interrupters, and are potentially more dangerous," Jamie smiles, "so we'll need to get some of those."

They load up with a hair dryer, a clock radio, lights, hair dryer, a clothing iron and many other items.

Says Peter Rees: "Sometimes it's hard to remember where each story came from but our main sources are obviously the things that are happening in contemporary news. There are a bunch of websites around that discuss this stuff. We have our own fansite, **www.mythbustersfanclub.com** , and if some folklorist is not writing a PhD on that fansite they should be.

"Everyone wants to read the books Jan Harold Brunvand puts out about urban legends – he writes these things up in a very scholarly way. But I think we've probably created a few of our own. As a collection point for raw, weird stories, it's unsurpassed. Even though we're not the highest-rated show on the Discovery, normally we have 4-10 times the web-traffic of any other show simply because people are fascinated discussing this kind of stuff. So those stories can come from anywhere, and the appliances in the bath myth has probably always been on the cards simply because we've all been told that by our parents.

"Whenever an engineer builds something that's over-complicated, that's the mark of an inferior designer. Simple is always better." – Jamie.

"Our question was: 'How do you actually test that without killing someone?' I don't know anyone who's actually been electrocuted in a bath but it's always in horror movies. This was actually one of the more dangerous experiments we did, in the phase where we were experimenting with having a safety officer.

"In the end it was unanimously agreed that having a safety officer was much more dangerous than anything in the entire Mythbusters safety history – first of all because he left the gunpowder on set one day, even though he was supposed to be in charge of it, it could have gone off, blown out half the set and burned the building down.

"Then he left the electrical current on in the bathtub, we'd been meticulous staying way away from this bathtub, we'd build a kind of small pool-like thing which we could actually put the bathtub in, so any water spillage we'd be separated from. He gave us the all clear to go on set, and Christine, the Mythtern at the time, put her hand in to drain the bathtub and got electrocuted because the guy hadn't turned the electrical circuit off. Fortunately it was a mild shock, but we find that Safety Officers are far more dangerous than if we handle it ourselves."

Bubble, bubble ... seriously, just don't try this at home.

The Mythbusters have widely different ideas of what will happen when these pre-loved appliances hit the water. The question is, "Would it kill you?"

Jamie says no.

"The analogy is like lowering a light bulb into the water after you've put the toaster in and expecting the light bulb to light up. Now I know that's not going to happen. Do I know that you're not going to get electrocuted? I don't think so but I'm not so sure I'm ready to jump into the tub just yet."

Adam says yes.

"Jamie doesn't realise that the live power of the AC is going to come through one leg of the bathtub and, with the drainpipe, is going to ground to earth and you will be a very fabulous conductor of the electricity in that short term the electricity is moving from the appliance to the ground. And it will kill you very quickly, I believe."

He adds, "To be honest, if Jamie doesn't think anything is going to happen, I have no problem in dropping a hair dryer in the bathtub while he's sitting in it. But something tells me he won't."

According to Adam, amperage – not voltage – is the killer, and the lethal limit he says is 70 milliamps. "Pump somebody with a million volts and it's going to hurt them, but if it's under 70 milliamps you're not going to kill them," he explains.

THE SET-UP

Having purchased just about every possible appliance you might drop in a bath, and a few more, Adam doesn't want to make the test "in any old bathroom"; he'd like an isolated bathroom where they can safely get clean tests. He gets the Mythbuster's building team (Tory and Christine) to build a bathroom set that is fully isolated, insulated and waterproof with a grounded bathtub above which is a re-setable collapsing shelf on which they can position the appliance before dropping it into the bath.

Mini Myths

Toothbrush in the toilet. Can material in your toilet bowl aerate when flushed and waft onto your toothbrush? At the end of a month of tests the 24 brushes in the bathroom have low level traces of faecal coliform bacteria. And, surprisingly, it doesn't make any difference how near or far they are from the toilet.

They create a white room, with linoleum floor – bath left, sink in the middle – complete with every detail down to the bin, toiletries, toilet brush and, of course, Adam's gown. They'll need to know how much current is flowing through the water, so Tory fixes wires to the drain under the bath which connect with the amp-reader, which in turn will produce readings for Jamie on his laptop computer. The bathtub is insulated and ready to go, now they need someone to get in it.

The Body: The basic idea is to put a body in the bath. In this case a body made from ballistics gel with two copper paddles in the dummy's chest, which represent the heart. Ballistics gel has similar properties as human flesh and should conduct electricity the same way. The paddles are to be wired up to a relay which will trip when the current passing between them hits 70 milliamps: ***that means death has been achieved.***

The test dummy gave a whole new meaning to bath gel.

Adam: *"Let's get down on our knees and pray. I don't know who to – is there a patron saint of ballistics gel? This relay is set to be tripped at about 6 volts, which we know from experimentation means that there's about 70 milliamps travelling across the heart.'*

"It's time to unveil our bath-bound dummy. But they wouldn't be the Mythbusters without pyrotechnics, so Adam has rigged up a flashlight which will ignite when the current hits 70 milliamps.

Adam: *"A lethal voltage across the heart of the dummy is going to short this wire – poof! – and we'll all know."*

"Let's get down on our knees and pray. I don't know who to – is there a patron saint of ballistics gel? – Adam

THE SCREW-UP

The test involves positioning various appliances on the shelf above the bath, connecting the reading to Jamie's computer, switching on the appliance … 3 ... 2 ... 1 … then pulling away the shelf so the appliance drops into the bath.

- **The hair dryer *(with a ground fault interrupter).*** They test the hair dryer in their workshop. This hair dryer has a ground fault interrupter, so if the Mythbusters have created a proper circuit it should cut out the moment it hits the water. And it doesn't. In fact, it pumps water and creates bubbles. The relay didn't trip and the ammeter didn't show enough amperage to kill.
- **The radio *(without a ground fault interrupter).*** "Okay," says Adam, "give me a countdown baby." Jamie: "And 5 … 4 … 3 … 2 … 1 … drop! Tory's shell collapses like a gallows trapdoor. And Jamie's laptop reads one milliamp, one 70th of a lethal dose of electricity. Something's wrong – the question is – what? The Mythbusters call in some electrical engineering know-how for a pow-wow. It turns out there are two faults in the system: *Adam and Jamie.*

Adam: *"Well, Jamie, this is just a classic Mythbusters screw-up. First up, the resistance of the human body in the bath is more like 1000 ohms rather than the 20,000 ohms that we read on the ballistics gel. Then it turns out that electrocution is classified as 6 milliamps across the heart, not the 60 we originally testing for."*

Jamie: *"And then I screwed up because I had the multi-meter on the wrong setting and I was measuring voltage instead of amperage, which meant we weren't getting the circuit grounded.*

Talk about getting your wires crossed. Oh well, even the Mythbusters make mistakes.

WATER WORKS

Adam and Jamie are trying to find what happens when an appliance splashes into your bath. When that happens the electrical current leaking from it is drawn towards the metal drain. H2O by itself isn't a great conductor, but the presence of salt helps. And because the human body contains a high concentration of salt water the current passes right through.

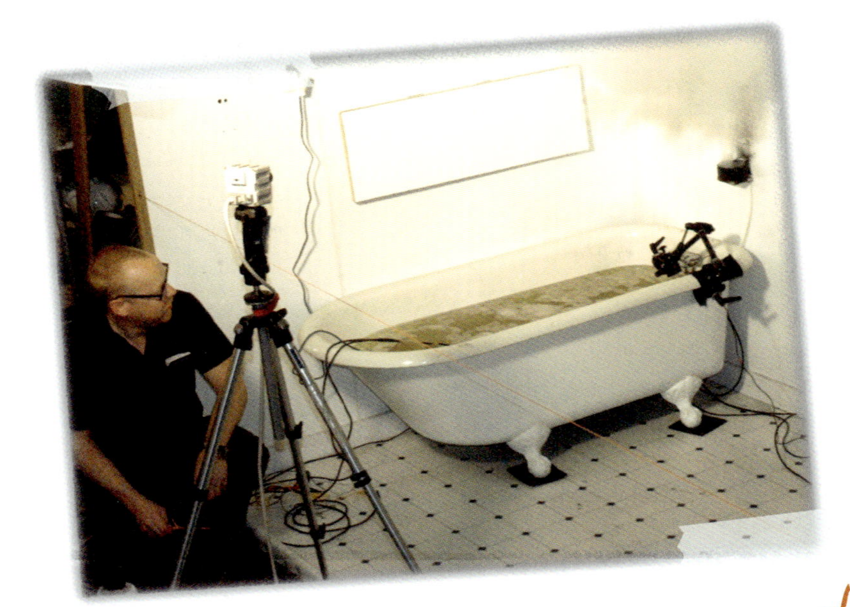

THE EXPERIMENT #1

The guys quickly re-calibrate the system and the GFI dryer takes another dive.

- **Hair dryer *II*:** The ground fault interrupter trips immediately, proving itself a life-saving device. This establishes that the circuit is now working. This older hairdryer doesn't have a ground fault interrupter on each drop. The amount of current across the panels is relayed to Jamie's computer. Adam sees 8½ milliamps across the heart. Jamie describes what would have happened to a human in that bath: "This one is dead as a dodo, well over the threshold". Just 6 milliamps fizzing through a body is enough to cause loss of muscular control, making it impossible to get out of the bath.
- **Toaster:** "Smell the danger. Okay, pulling in **3 … 2 … 1 … SPLASH.** Our friend here is toast," says Jamie. "At 12 milliamps, that's totally lethal."
- **Curling iron:** "Let's go! **3 … 2 … 1… SPLASH.** It's lethal at 6 milliamps, but this reading of 4.5 milliamps is a too close for comfort. I think I wouldn't want to be in that water," Jamie notes.

- **Clothing iron.** And now the iron. **SPLASH. FLASH!**

Adam: *"Wow, that's the highest spike yet, we got 32 milliamps, this thing is literally pumping death into our subject here. And 32 milliamps is more than five times the lethal limit."*

CONCLUSION:

Death. Says Adam, "Items that have large heating elements are delivering a lot more current because they have a lot of exposure to the water."

The Mythbusters also discovered that if the appliance lands closer to the metal drain, the dose of electricity is drastically reduced because the current has less distance to travel.

THE EXPERIMENT #2

Having established that you can be electrocuted, the Mythbusters want to find out whether the water's conductivity is changed by additives to the bathwater.

- **Bubble Bath:** Checking his laptop Jamie says, "It's only about 11 milliamps, last time it was 18. I believe the soap suds cut it down. So bubble bath might save your life."
- **Urine:** Next step, the Mythbusters try 200mls of fake urine in the bath. *(Urine is just water with salt and food dye added.)* Poof of smoke. Adam: "Fake wee test: **3 ... 2 ... 1... SPLASH.** Jamie: "That's got it, it's just in the death zone." So urine kills.
- **Epson Salts:** Salt normally increases water's conductivity. Splash/Flash. "Wow!" exclaims Adam. Jamie says, "That was huge – you'd be toast. You'd fry. You'd boil." (This from the guy who thought nothing would happen.)

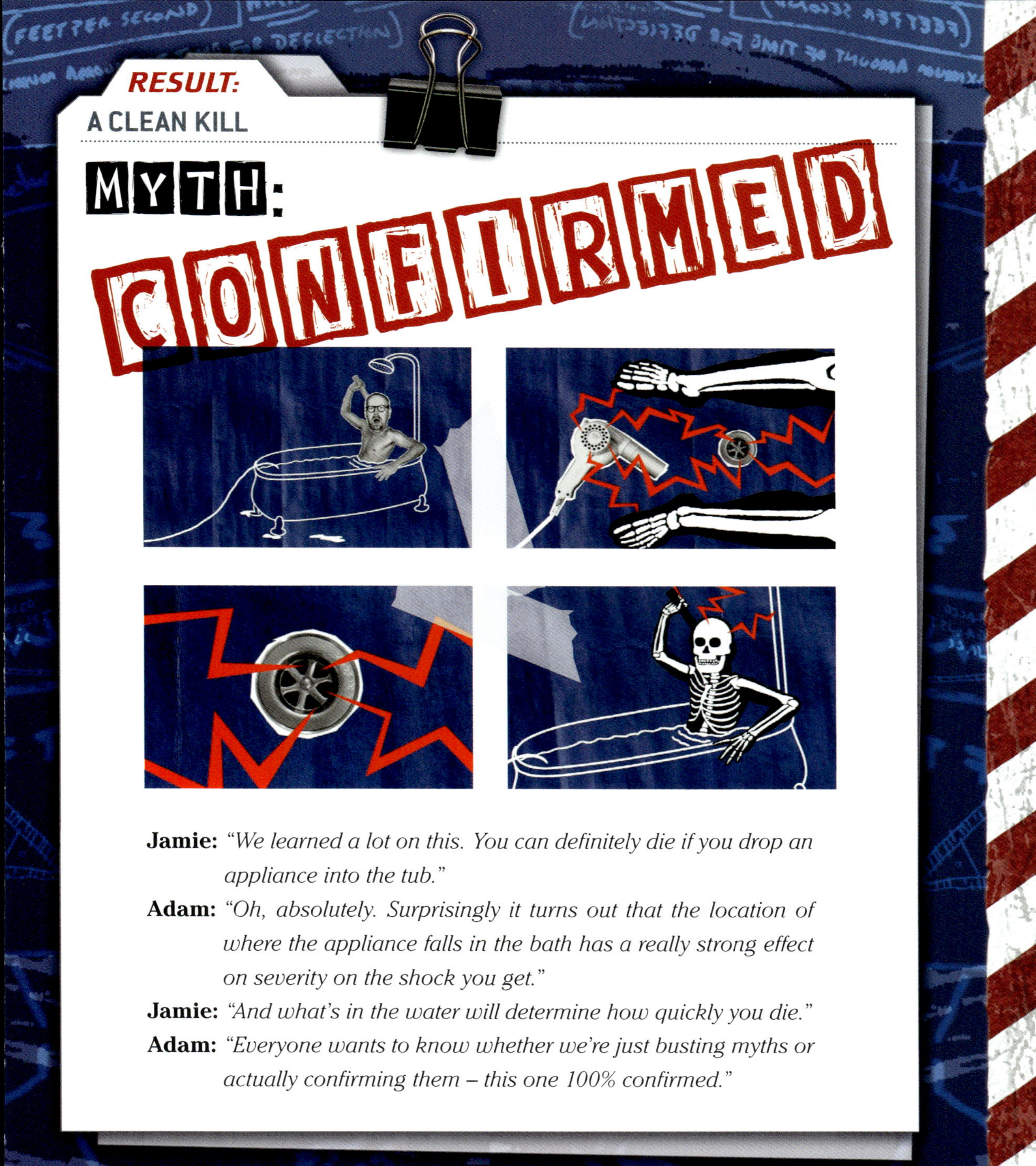

RESULT:

A CLEAN KILL

MYTH: CONFIRMED

Jamie: *"We learned a lot on this. You can definitely die if you drop an appliance into the tub."*

Adam: *"Oh, absolutely. Surprisingly it turns out that the location of where the appliance falls in the bath has a really strong effect on severity on the shock you get."*

Jamie: *"And what's in the water will determine how quickly you die."*

Adam: *"Everyone wants to know whether we're just busting myths or actually confirming them – this one 100% confirmed."*

M = MASS (SLUGS - 1 SLUG =
MAIN VARIABLES
V = VELOCITY (FEET PER SECOND)
T = TIME (MAXIMUM AMOUNT OF TIME FOR DEFLECTION)
CRACK

Think Quick

CAN QUICKSAND KILL YOU?

The Myth: ***We know that quicksand not only exists, but it can pull you down and kill you?***

It's another classic Hollywood horror story: movie star Johnny Weissmuller could always rely on killer quicksand to swallow Tarzan's enemies. But is there even a grain of truth in the killer quicksand story?

While bushwalking in Australia, Peter Rees has been up to his waist in quicksand at the bottom of the Goulburn River. He says, "You can also do it on Fraser Island, and on Rainbow Beach there's actually places where natural springs come up through the sand – it's basically water that's got a lot of sand in it, and you can bob around."

So the truth is, before conducting this experiment the Mythbusters knew the myth wasn't true.

"We make this stuff up as we go and it's always surprising when it actually works like we thought it would." – Jamie

"But Jamie and Adam didn't know that," says Peter. "And that was a major build, and that's the kind of thing Jamie and Adam are fantastic at. You say to them, 'How the hell can you build something to replicate natural quicksand?'

"I wouldn't have the first clue, but with the right combination of pumps and a recycled sewerage container they did it. There were problems with it, it kept clogging up, but all this kind of thing is where they excel. Give them that task and they will absolutely pull it together and come up with something that will work in the end."

I didn't know Adam had a licence to drive heavy earthmoving equipment?

"He doesn't. In fact, one little-known thing in Mythbusters is that no-one has any qualifications in anything."

Except Scottie.

"Scottie has welding qualifications. Oh, and Grant is an electrical engineer," adds Peter. "I've got qualifications in environmental management and English literature, but Jamie and Adam are completely self-taught, or taught on the job. So none of them have qualifications and that works very well because Jamie is very fond of saying that his job is using tools for purposes for which they were never intended."

Quicksand is real. Just ask quicksand survivor Gerald Gray. "When I was eight years old I was on horseback riding with my family on the St Anna River in California," Gerald says. "This is a very sandy-bottomed river, we knew it had quicksand but we never could tell where it was going to be.

"Their horses missed it and got to shore but my horse went down. He sank to his

Mini Myths

CD ROM Shattering. *The myth goes that some CDs shatter in very high speed drives. Reports say the physical limits of a CD have been reached when pushed into the 30,000 revs per minute zone. Is this true? The Mythbusters reckon it's not.*

belly and I was afraid to get off and I was warned not to get off because there was quicksand between me and the shore, but they managed to get ropes around my horse and their horses pulled him out."

The question isn't whether quicksand exists; it's real all right. The question is does *killer quicksand* exist?

Time for your close-up ...

THE SET-UP

Quicksand is the result of a process called liquefaction. Water rising from under the ground over-saturates the sand or soil, reducing the friction between the particles until they can no longer bear any weight. The Mythbusters are going to attempt to create a killer quicksand that has some kind of suction effect that would drag someone all the way under.

To do this Jamie and Adam need certain ingredients, which they find over San Francisco's Golden Gate Bridge at the Ceramics & Crafts Supply Company. Jamie believes there has to be a threshold of fineness in quicksand where at a certain point a person would not sink but stay in suspension, as they would in mud. And to create this Jamie and Adam need the lowest density sand they can buy.

"A bag of your finest sand please, and we do mean your *finest*," says Adam. The guys settle on two types of fine-grained sand and some clay powder. "If we come up with some killer quicksand, we'll give you the recipe so you can keep your patrons happy," he jokes. The shopkeeper retorts: "I probably have customers who would like to have it. So that'll be great!"

Adam and Jamie need to know the density of their quicksand mix. If it is more dense than normal water, it might be a little difficult to sink, so Mythbuilder Christine is sent off to make a hydrometer, an instrument which measures the liquid density. A hydrometer is set afloat with a calibrated stem, with the lead shot acting as ballast to keep the float vertical; she takes the reading where stem and liquid meet.

Because the ultimate test is going to involve a lot of materials, the Mythbusters

"You'll swim in the quicksand and enjoy it!" Jamie gives Adam a little encouragement.

want to first figure out what grades of sand work best. To do so they will use 20-litre buckets and some plumbing which enables water to rise through the bottom of the sand with some sort of filter device on the bottom. Once they figure out how it works, they'll know what they're going to do in the big picture by seeing if there's a difference between how the different sands behave in this small-scale flow of water.

Time for a small-scale test in a small bucket:

- **Regular sand:** The sand is totally compacted. When the water starts to rise, Jamie thinks the water needs to be diffused more evenly through the sand, so they drill more holes and add more gauze with a fabric which is a much finer weave and should hold their fine sand, silt or whatever they decide to submit to the bucket test.
- **Fine craftstore sand:** Jamie puts his foot in the bucket and concludes, "It's not like I'm plummeting to the bottom, but it is definitely letting me drop a lot quicker than sand normally would. And it also is resisting pulling back." Adam suggests that might be enough to cause a problem if your whole body is doing that.

- **Finer craftstore sand:** Again, Jamie puts his foot in the bucket and finds it even harder to retrieve his foot, lifting the bucket in his attempt. Adam concludes that the finer sand makes a difference in its powers of suction.
- **Clay powder:** Jamie dismisses it "This isn't working," he says. "This is just mud and it won't let the water flow through it at all."

The Mythbusters are trying to create quicksand. Not ordinary quicksand: "killer quicksand", stuff that will suck you right under. Says Adam, "Without a doubt we've got this wired, I think we're going to be able to dial the flow in. It's a lot of stuff to move around, like 20,000 pounds (nine tonnes) of sand and thousands of gallons of water, big pumps and stuff. It's a helluva deal!"

OBSERVATIONS

Jamie: *Our task is not only to build quicksand, but killer quicksand. I have my doubts that we'll be able to do this because the sand actually makes it harder for you to drown in it because it is, in effect, making the water more dense.*

Adam: *I'm pretty sure it'll work, it's just a question of how well. While I know we're going to sink in there, I don't know how fast we're going to get stuck.*

THE EQUIPMENT

For this one the guys are going need one mighty big bucket – a 7500-litre tank (that has previously played a starring role before on Mythbusters) which Adam cuts in half from the inside, to a bit over two metres. Adam then drills the round holes a hand span apart, and explains, "These gauze holes will allow the water to get out while keeping the sand inside, and that'll make the top of our quicksand not like a slurry, but to actually resemble real quicksand." Meanwhile, Jamie and Christine start work on the diffuser. Christine is making the gauze hole-stoppers. The gauze will let water rise into the tank, but restrain the sand from falling back into the pump.

"One of the things we do here for a living is normally using tools and materials in ways they are not intended to be used." – Jamie

The Plan: Adam explains what will happen next, "We've finally got this huge tank about eight feet high high. We're going to feed water up through this diffuser into seven feet of sand, and when we get that water to flow at the right rate – what we should end up with is quicksand." The tank is positioned in a plastic pool, which serves as the catch basin and reciprocating tank for their quicksand bay, and the water is turned on. Next, nine tonnes of sand is trucked in along with Tom Holtzer, an engineering geologist.

Last check: Adam checks the feasibility of the experiment, "So, Tom, this sand we've got here, is this great for quicksand?"

Tom: "It's as good as it gets. Very fine-grained, there's so much surface area here which means when the water flows it can grab onto that and that'll create the quick condition."

This irrigation pump can move up to 5000 litres a minute but if it stops while Adam and Jamie are in the quicksand, the sand will suddenly compact and the Mythbusters will have some pressing problems. Adam and Jamie are suddenly facing danger on two fronts: on one hand the guys could be sucked under by killer quicksand, or they could be crushed. The killer quicksand rig is almost ready. The bottom of the tank is lowered into position and, driving some heavy earthmoving equipment, Adam pours the sand. Says Jamie: "We make this stuff up as we go and it's always surprising when it actually works like we thought it would, and that's kinda the way it looks like it's doing. Everything is coming together relatively quickly."

Adam then gets in. The sand is solid. "What do you think?" asks Jamie.

Adam is ready for action. "Let's turn on the pump and start getting some hydrations to the centre," he says. This Jamie does, to which Adam responds, "Unfortunately, our ports aren't letting out the water nearly fast enough. These ports are supposed to let water out before it reaches the top of the sand so it can look like a sandy beach. Unfortunately the sand is so fine it's clogging up those ports and the water is just going up and over it." The rig is working, everything is working exactly as planned, they just need to filter the sand a little better to get something that looks good. This they do using some cotton batting which will allow more water to flow through while still holding back the sand.

Meanwhile, Christine has managed to finish the hydrometer. She says, "This hydrometer will give us the reading on this water and will tell us if we put it in the quicksand whether it is more or less dense than this water."

The hydrometer floats in the water, its buoyancy level is marked, then it is dunked in the tank. Says Adam, "It's floating, and shows really nicely that the water-sand mixture is a lot more dense than water."

So, in theory, a person will float better in quicksand than normal water. The killer quicksand myth is on shaky ground.

THE EXPERIMENT

Says Adam, "I'm pointing out quite realistically that if the pump stops during the experiment we could die because we'll be encased in 20,000lbs of sand, which is cause for real concern."

Jamie: *"Worse case scenario is that Adam dies. Of course we don't want to see that happen, I'd hate to have to hunt for another co-host. But we've got all the precautions that*

The quicksand myth demanded one of the more elaborate setups.

we think are reasonable to have on this and I'm fairly sure that he'll be okay one way or another."

Adam: *"We have several escape techniques, first and foremost we have whatever one would want if they were near quicksand: a vine. Second, we have a walking stick which is often recommended in the literature we have read as a helpful tool for getting out of quicksand.*

"And third is a bar which we can throw across the top of the tank and I should be able to hold on to in case I get really desperate. At that point, I'll be making those sounds that you just don't make when you're kidding."

One more accessory: the pith helmet is always a good technique for finding someone who has drowned in quicksand because it will be left floating on the surface.

1. The Adam Savage Test: Adam climbs up and lies in it. Mimicking the movies Adam says, "Oh look it's a white sandy beach, I don't feel in any danger at all." You will Adam … you will. The pump is switched on, water begins to rise through the sand. As liquefaction occurs Adam gets that sinking feeling falling first to his knees, then to his chest. "It's more dangerous now than I thought," he says. "It's good going down really slowly, when I sank to my chest in the sand and water mixture I thought I'd got to my buoyant point.

Now we turn the pump off. "I tell you I'm not going to be able to get out of here until we turn the pump back on." Adam is far less dense than the quicksand and so he's suspended. Jamie presses him down, he bounces back up. Adam concludes: "I think we've created a very realistic scenario for our worst case scenario quicksand, using incredibly fine sand, usually called sugar sand, we've got water bubbling up through it like an artesian well and when we get it working we definitely go down in it." The vine is great. Adam pulls himself out.

2. The Jamie Hyneman Test: Now it's Jamie's turn in the tank and he also sinks to his chest. Enjoying the moment, Adam tries pushing him deeper. "Down you bastard! Down!" Jamie concludes: "It's amazing that it's so buoyant, I'm totally floating here. And I'm not touching the bottom so look at the buoyancy on that!"

RESULT:
THINK QUICK

MYTH: BUSTED

WHY?

- The Mythbusters know that people and animals have died in quicksand, but believe that as they were incapable of extracting themselves they stayed there until they died, probably due to exposure or dehydration.
- They do not accept that the quicksand slowly dragged them under and then drowned them.

Jamie: *In fact, they're quite buoyant because the quicksand is a mixture of sand and water that is actually quite heavy and you float like a cork on it. You're going to bob around fairly high up in it. The best thing you can do is lay on your back and use the vine, walking stick or whatever happens to be handy to try to gently pull yourself out."*

So is the Mythbusters' final word on Hollywood-style killer quicksand is confirmed, plausible or busted?

Jamie: *"I'm surprised you even have to ask, it's absolutely busted. There is no such thing as killer quicksand."*

Adam: *"I agree, completely busted."*

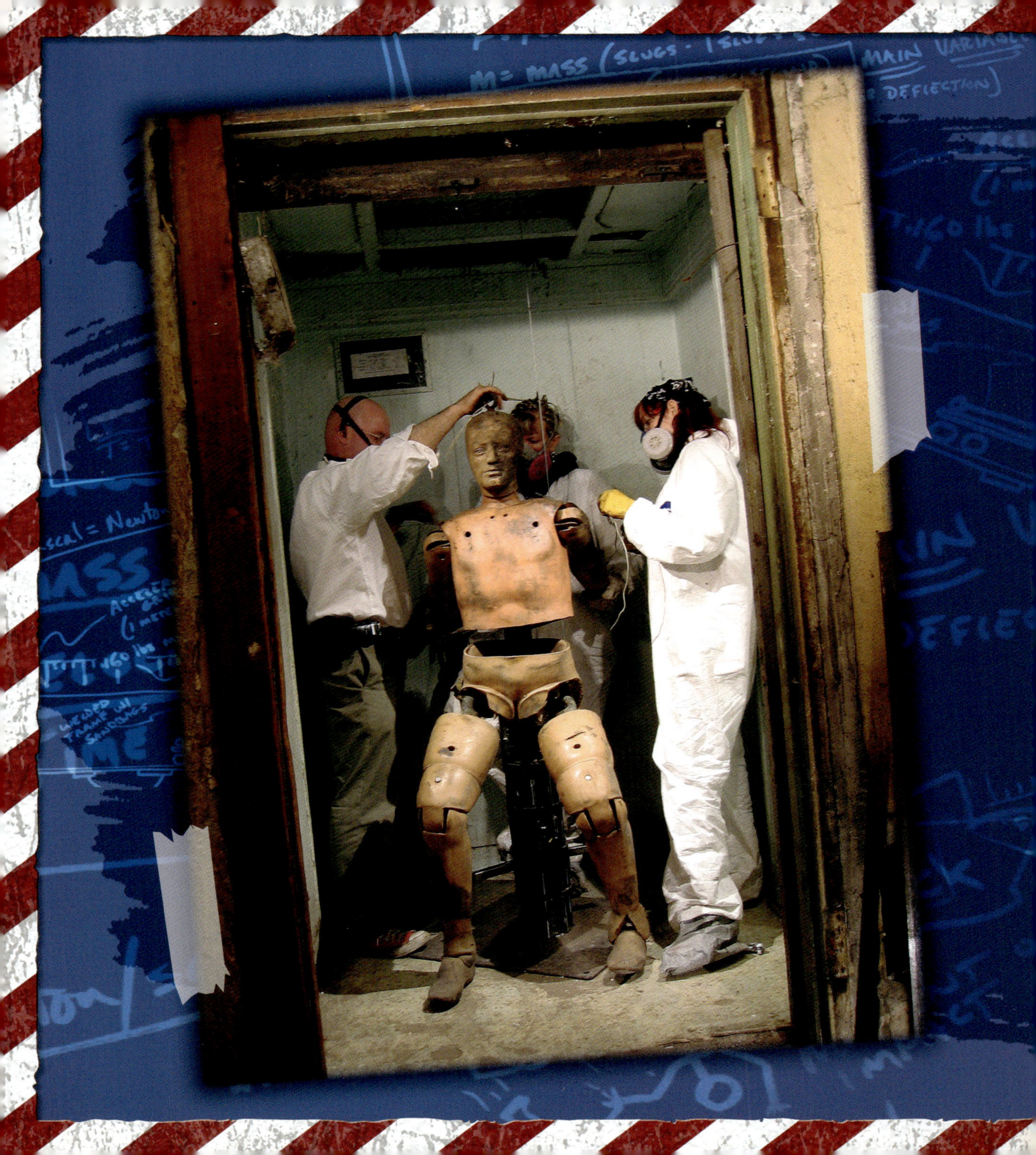

Going Down?

ELEVATOR OF DEATH

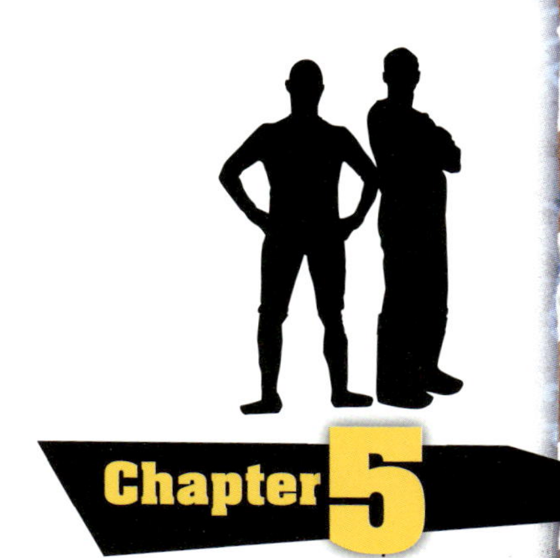

Chapter 5

The Myth: ***You're in an elevator, the cables are cut, you're plummeting rapidly towards the bottom of the elevator shaft and right at the very last second you jump as hard as you can – and that saves your life.***

In July, 1945, a terrible tragedy at the Empire State Building gave rise to a miraculous survival story: an elevator plunged down 80 floors, cables and debris tumbling over it.

It was a drop of more than 300 metres. Inside was a young elevator operator who was found alive in the basement by Don Maloney, a heroic coastguardsman. He said, "The guard inside was badly crushed and burned. I got her out, gave her morphine and tended her wounds best I knew how." How did she survive that crash?

The question is – what would the Mythbusters have done if this test hadn't worked first go? Believe it or not, they were lined up for a second take. They were lucky it worked, because had they failed they were all set up with all the welding equipment and whatever was needed to get it rolling again.

"We have been incredibly lucky throughout our entire series, that in a lot of the

really big things have worked first go," says Peter Rees. "Funnily enough the big things seem to work more reliably than the little things."

The behind-the-scenes battle in this myth was between the pragmatic researchers and the people who wanted the show to look good for television. Frankly, they could have built a mock-up of an elevator and dropped it from a crane – that would have tested the myth, and scientifically it would be exactly the same thing as using a real elevator. And so the two departments argued for three months; the "creatives" argued: "It may well be the same thing in terms of gravity, the physical space on the elevator the actual falling, but our viewers expect a full-scale test!"

Second, who knows whether the elevator might not act like a plunger inside the shaft, and the air that's underneath it might actually break it? Who knows whether on each successive fall the doors release enough air so that it accelerates at its full velocity? Who knows whether a crane can drop a lift perfectly square? There was only one way to find out – use a real lift.

As it turned out, the biggest battle in that story was finding someone who was willing to let them drop their elevator.

The theory behind this myth proposes that being in mid-air at point of impact disengages you from the catastrophe. Lester Appel, of Apple Elevator Co, has worked on elevators for 30 years and he's not falling for this myth. "If you're in an elevator you're travelling at the same velocity as the elevator itself," he argues, "and you're not going to overcome that physical fact."

What would Lester do if he was trapped inside a falling elevator?

"I believe if I had presence of mind, I would try to lay myself down on the floor – assuming I could get there - just to minimise the injury to my body."

THE SET-UP

The Mythbusters will need to drop an elevator down a shaft, and just before it hits bottom the test subject will jump in the air. They'll measure his speed to see if he's avoided what seems certain death.

There are two elements in this test. First, the Mythbusters need to locate an elevator that they can drop from a great height; second, they need to find someone who is willing to get in it, and there's only one person on the team who can't argue – Buster, the longsuffering crash test dummy.

Destruction seems assured, and Jamie can't wait to scout a location. There's a fire-training facility which sounds promising in Pleasanton, outside San Francisco. It's quite a complex with an elevator which drops five floors, which is 12.1 metres. That's an impressive drop. But for this myth, they're looking for something really special. Jamie's got a lead on a derelict hotel in nearby Oakland that's eight stories high – and after checking it out with builder Kari, they feel it will do the job.

With an old elevator weighing more than a tonne to consider, the boys took extra precautions.

The Grand Old Hotel Royal was built around 1910 but she ain't so grand any more and scheduled for demolition. They find the elevator, but it's not going anywhere fast. "There's the motor, what's left of it," Jamie sighs despairingly. "And that's the elevator, it feels like steel, but I don't know if it'll stand up to an eight-storey drop without crushing."

Getting the elevator working will be no small task; the cables haven't moved in years, and the motor is a wreck. But Jamie is hopeful and says to Kari: "The cables and everything are here, it's a little rusty. If we could tie power onto that we would have a working elevator. If we can get it to work it's going to be perfect."

Next, Jamie and Kari climb the staircase all the way to the top. It's a long way to the top of the shaft, eight floors (nine counting the basement). Jamie gets out his tape measure and announces, "92 feet (28 metres) to the top of the pulleys."

Kari adds: "92 feet to the dead bird!"

Yes, the place is filthy. It's nasty enough in the Hotel Royal to wear dust masks and Jamie and Kari have to clean up the filth: there's pigeon poop, broken fittings, dust, dead birds, busted tiles and everything you'd expected to find in a derelict construction, maybe even ghosts. "Don't leave me here alone!" cries Kari, scampering along the staircase after Jamie. Even Jamie finds the premises "kinda scary".

A 28-metre drop, 680kg of elevator car … it looks like they have their location, but have the Mythbusters bitten off more than they can chew?

TWO CHALLENGES

This experiment has two engineering challenges before the drop:

1. First to figure out a way to raise the elevator.
2. Second – to rig the victim to jump up at just the right moment.

Pleased with the location, Adam's first role is to find all the necessary rigging equipment to pull the elevator to the top it. To Jamie he assigns "the hard job" - getting Buster to jump at the last second.

Raising the elevator

Meanwhile, Adam is looking for rigging, asking all kinds of questions salesmen don't hear too often, like: "If you cut this cable and attach basically this quick release between the cables and my elevator so that I can then drop the elevator – what would you recommend?"

Mini Myths

Goldfish memory myth. *Goldfish have a three-second memory – meaning as far as they are concerned, every trip around a round goldfish bowl is like a new experience. To establish whether or not goldfish have memories the Mythbusters have to teach them something, which they do – concluding that goldfish do have memories, and are probably bored living a life within goldfish bowls.*

Rigging salesman George Yuan helps Adam settle on extra-strength cables and a quick-release double strength shackle. The whole experiment, and invariably the Mythbusters' safety, hang from this quick release shackle. Everything has to be right.

Jamie's meticulous preparation was never more important than during the elevator test.

Jamie has now got to get the elevator to go up. It's awfully stiff, but it's moving. Jamie says, "Movement means improvement". He has an idea to modify an electric vehicle he has built – the same electric vehicle used in the 7-Up commercial but extensively modified. He calls it the Life And Death machine, or LAD for short. Says Jamie: "I built a purpose platform that has the electric motors, the gear reduction, speed controllers, electronics and batteries all compacted into as tight a unit as I could come up with. Then we built these treads on it." Jamie sees the LAD having applications in fire-fighting and law enforcement. Today it's got to lift an elevator. He wheels it into the building, smashes a wall to get it to reach the elevator. Remember, this elevator hasn't budged in years. "It's moving," he cries. The gears turn manually, the wheels spin slowly.

Amongst the dust and the echoes at last the elevator begins its slow climb with cheers as it reaches the second floor, because having achieved that they know it will go the rest of the distance – dangerously perhaps, but it will do it.

Adam: "I think Jamie's rig is classic Mythbusters. We have to lift an elevator and Jamie says, 'I've got a robot in the shop that I can modify to do the job'."

They secure the car at the top.

"People seem to be calling us a lot nowadays when they want bad things to happen to their stuff." – Jamie

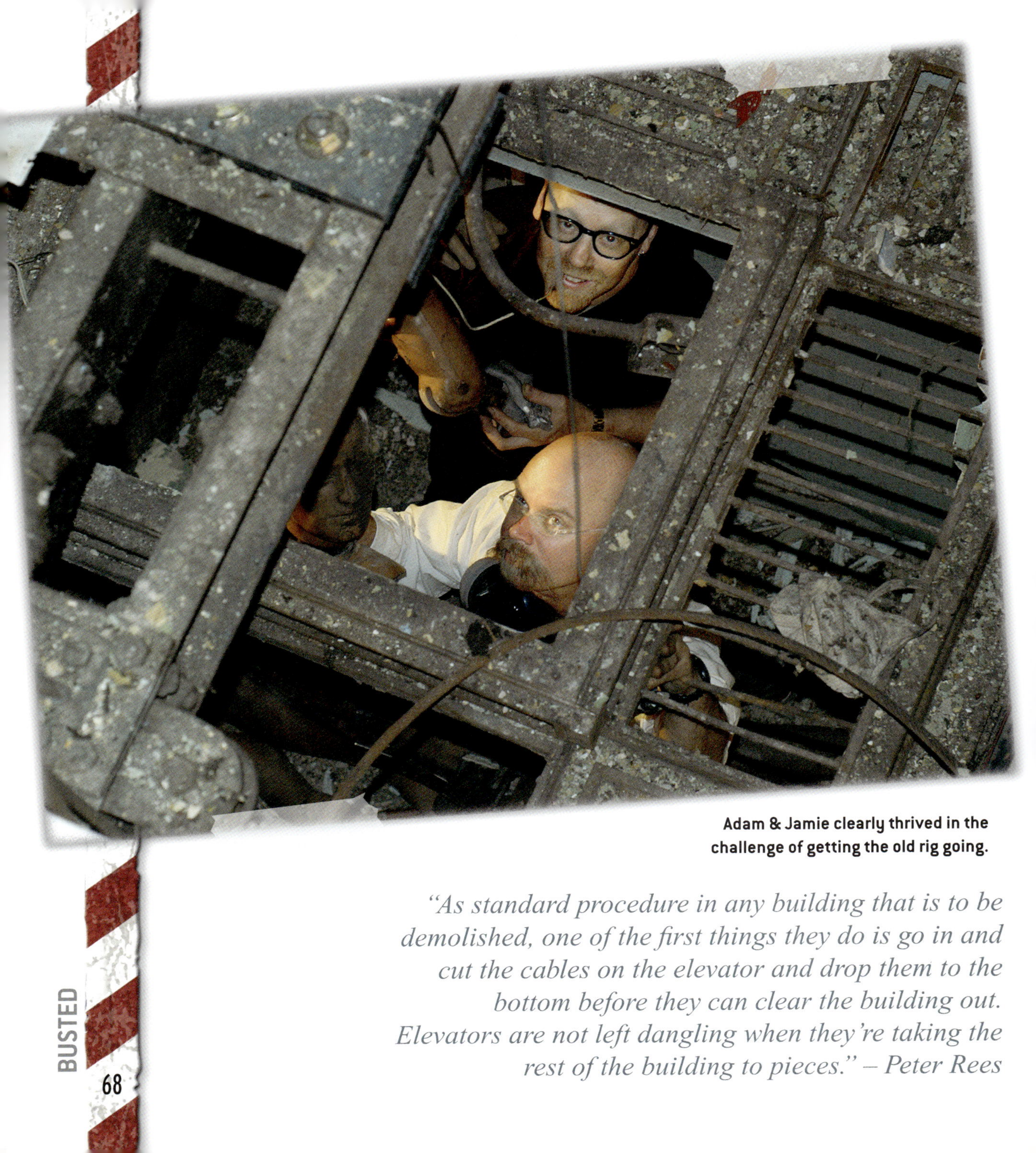

Adam & Jamie clearly thrived in the challenge of getting the old rig going.

"As standard procedure in any building that is to be demolished, one of the first things they do is go in and cut the cables on the elevator and drop them to the bottom before they can clear the building out. Elevators are not left dangling when they're taking the rest of the building to pieces." – Peter Rees

Getting Buster to jump

Back at the workshop, Jamie, Kari and Scottie are dissecting the human jump by jumping themselves. They conclude that in the short moment of acceleration the average top speed generated is just over 2.4 metres per second. They need the victim to replicate that.

"I have to figure out how to make Buster jump," Jamie says. "I think I'm going to have to use some kind of spring because I don't want to go for gunpowder." No gunpowder is good news for Buster; instead, they plan to use a kind of turbo-charged pogo-stick, a car spring inside overlapping tubes with a quick release catch. They'll rig Buster to sit on top. Scottie's specialty is working with metal, so in no time she's back, having created Buster's launcher out of metal tubing. The spring is tensioned, and the quick release set. Then Buster is lowered into place. Although it's a powerful spring, when Jamie launches Buster it's not much of a jump. Attempt one was more a jiggle than a jump.

"I believe this is more damage than we've ever done to Buster in one myth before." – Adam.

To reduce the pogo stick's weight, Scottie cuts out as much of the heavy metal cylinder as possible and adds a second spring for extra power. She also smears the whole assembly with lard to cut friction. The result of the second jump is better, but still not near enough to replicate a human jump.

The real problem is power. Buster needs more spring in his step. If that spring is going to turn up anywhere, it'll be in a car yard. And Scottie finds it.

Scottie: "Did I do something good today?"

Jamie: "You did great. Thank you. Oh yeah, it fits!"

Buster may at last be ready for his great leap forward. After the necessary boom, the third jump is a success. Spirits are soaring, Buster reaches a maximum velocity of nearly 3.8 metres per second, compared to Jamie's estimated average top human jump of 2.4 metres per second. Jamie: "That was significantly more than we got before – it looks like a good jump!"

They've raised the elevator, got Buster to jump and have solved half the elevator of death problem – getting the Hotel Royal's elevator working again lies ahead.

THE EXPERIMENT

Elevators are designed not to fall. Fortunately the Mythbusters know how to dismantle safety systems, beginning with the emergency brake. After a fiddle – during which the team is understandably tense – Adam emerges holding up a small piece of metal and, with a smile, he says, "Anticlimactically, I believe I've disabled the entire mechanism by removing this single pin!"

But they're not in freefall mode yet. As well as being linked to a pin, elevators are linked to a counterweight which weighs about the same as the car when half full which also has to be disengaged.

And now it's time for the great elevator drop. Can Buster survive a 28-metre drop if he jumps about four metres up at the last second? Jamie and the team are in the basement, Adam is up top with his finger on the quick release shackle.

Says Jamie: "We don't want to do anything without being absolutely sure of what's going on. It's nasty, someone could die here." Says Adam: "I'm terrified, I'm totally excited. Ready, drop in 5 … 4 … 3 … 2 … drop!"

CRASH!

Eight seconds later, after an enormous blast of air, dust and a resounding crash Adam emerges laughing, "That was intense!" Laughter all round, but it's time to see how Buster fared. The overhead camera shows Buster's pogo stick launched right on cue, but did it save him?

Well, he still has his head. But his foot has come off, his leg is broken, his other leg twisted, left arm attachment and shoulder blade broken, ball joint missing from his waist, "He's still smiling though …" says Scottie.

Adam sighs, "I believe this is more damage than we've ever done to Buster in one myth before."

RESULT:

ELEVATOR OF DEATH

MYTH: BUSTED

WHY?

By the time it reached the bottom of the shaft the elevator and Buster were falling at 85kmh. Buster jumped up at a metre per second, or 3200 metres an hour – enough to hit the roof – but he still hit the ground at a fatal 82kmh.

Says Adam, "I think Buster feels good about a job well done and he just hopes that we don't put his head on backwards." Buster will be repaired, but if he's smashed to pieces how did that elevator operator survive her plunge in the Empire State Building from 10 times the height?"

It turns out the car was a tight fit inside the shaft, so trapped a cushion of air beneath it. And the cable hanging below coiled into a kind of spring as it fell. Those two factors slowed the car enough to save her life. Alas, fate was not so kind to Buster.

Jamie: *"Well, you might be weightless relative to the elevator, but you and the elevator are going down at that speed. So unless you can jump up at the same speed, you're dead."*

Adam: *"I think we've done all that we can here, I think this myth is definitively and finally busted."*

Jamie: *"It was fun though!"*

M = MASS
V = VELOCITY (FEET PER SECOND)
MAXIMUM AMOUNT OF TIME FOR DEFLECTION
Seagate

A Tissue Issue

Chapter 6

CAN A FLYING TISSUE BOX KILL YOU?

The Myth: ***A tissue box sitting on the back shelf of your car could gain enough velocity in a car crash to kill you.***

While leaving her local shopping centre in the United States a young woman involved in a minor fender bender was killed by the groceries she'd purchased. They flew off the back shelf of her car and hit her in the back of the head.

It may seem like a freak occurrence but, in 2001, an estimated 13,000 people were injured by unrestrained objects in the back of the car during an accident. In Australia, there were two cases of drivers being impaled by their golf umbrellas.

When Mythbusters and automobiles meet head-on, the result is usually carnage for the car. So what's on the menu today? One car, three crash barriers, a nervous-looking Buster and a tissue box.

The Mythbusters must have known the flying tissue box wouldn't kill, surely? Indeed, they must know the end from the beginning in quite a few.

"I've got to say the best stories are the ones where we don't know and no-one else does either," says Peter Rees. "My aim would always be that none of us know the outcome, but unfortunately that's not always the case, nor is it always the most interesting way to approach the stories. It actually had a positive spin-off because, although a tissue box can't kill you, there are a whole bunch of other things which people regularly put on the back dash in cars which definitely would kill you in an accident." This theory says that in an automobile accident a tissue box sitting on the back shelf of your car could be propelled forward with enough force to kill you when it hits your head. Is this a tissue of lies?

The most spectacular way to test this myth would be to place a tissue box on the rear shelf of a car, get Buster behind the wheel and crash the vehicle at high speed – which is exactly what the Mythbusters are going to do.

THE SET-UP

For the ***Killer Tissue Box Test*** a Plymouth Fury, already in their possession from a previous experiment, is the chosen sacrifice. As for the killer tissue box, the distance from the back shelf to the driver's head is just more than a metre. Anything that is moving forward is also going to be affected by gravity, so it's going to go down. However, there's generally a lip created by the backseat which is pretty universal in most cars. Jamie: "It's a nice little ski-jump."

The plan is to crash a full-sized car with a tissue box in it. Problem: the Mythbusters don't just want to test a tissue box – they want to test a variety of objects, but they don't have enough cars to crash one for every object they want to test. This requires them to build a repeatable crash test rig where they can see how each object reacts under the worst case crash conditions without destroying a rig every time they use it.

The rig: The rig will have to be lightweight, yet able to handle huge G-forces. Even at low speed collisions people can be subjected to around 80 Gs – astronauts pull only eight at lift-off. Jamie: "What I decided to do is to attach an aluminium rig to the underside of a sheet with reinforcement on it. Weight is actually critical with this thing because the heavier it is, the harder it's going to be to stop." The rig will be attached to the tow vehicle using a break-away hitch which should be strong enough to support the weight of the rig during acceleration but weak enough to break at the point of impact. Adam adds a rear shelf for a tissue box to sit on.

The crash test vehicle is ready, several ballistics gel boxes representing the human head have been made and it's time to take this show on the road. Adam and Jamie have found an unused air strip which suits the job because they need a long straight open run. They've got 300 metres of cable; one end is attached to the crash test vehicle, the other is attached to three concrete barriers each weighing 1800kg. The barriers will act as an anchor – when that cable is pulled straight their combined weight will bring the crash sled to a very sudden stop. They've hired a pick-up truck that will tow the sled up the runway at highway speed. Jamie: "The first thing I need to do is test how fast I can I get this car going over the distance of 300 metres because that is the length of cable we purchased for testing this myth." It gets to 120kmh. Adam makes some final adjustments to the crash test rig. "Well we've got a back-seat here and we've got the ballistics gel making a head there," he says.

Time for a test run. "This is when we find out if all our hard work has paid off," Jamie says. The truck reaches 100kmh; when the sledge hits the end of its tether it should stop dead, but the cable snaps and the crash sled is like a dog let off its leash.

Take two: they've got 90kgs of metal on the move at 100kmh: crash car and cable part company again. This time the cable ripped a half-metre hole in the aluminium I-beam, and it's also taken a sizeable chunk out of Jamie's ego.

"We could use some heavy nylon strapping and wrap it around that whole I-beam and maybe that'll distribute the load a little bit more," Jamie says.

Between those two things the Mythbusters should be able to get some good tests out of this rig. These new straps have a combined braking force of 18,000kgs. The crash-rig stopped right on cue. At 70kmh the cable stayed attached and the concrete barrier was dragged along like a piece of Styrofoam.

Finally they're ready to do some crash-testing with some objects you are likely to find on the back shelf of someone's car.

- **The figurine:** Although it weighs only 125 grams, small objects can pack up to 30 times their own weight in a sudden impact. Unfortunately on this run it simply fell over into the fence of screws and did not move. **Verdict:** Jamie: "At a certain speed I'm sure the bubble head could be lethal but I don't think that speed is 70kmh."
- **The sub-woofer:** The crash rig slams to a halt, the sub-woofer keeps on going. **Verdict:** At 70kmh you'd really be able to feel the beat.
- **The hatchet:** At 70kmh it'd be like something out of a B-grade slasher film. The axe head comes out on the other side of the board. **Verdict:** Jamie: "I think it's safe to say, don't put a hatchet on your rear shelf!"
- **The fire extinguisher:** A one-kilogram fire extinguisher probably wouldn't be fatal, but it would definitely ruin your day. **Verdict:** It'd give you concussion, you'd be hurting.
- **The bowling ball: Verdict:** It's like a cannonball. And remember, Jamie was only doing 70kmh.
- **The tissue box:** It's time to see what a tissue box does in the worst possible scenario.

Jamie: *"Okay, this is the 120kmh test - the maximum speed!"*

Adam: *"5 … 4 … 3 … 2 … 1!" Jamie ups the pace."* **Verdict:** *The box of tissues impacts but, at 323 grams, it just doesn't have enough mass to do much damage.* **Jamie:** *"The tissue box might be like 'Ow!' but we're guessing an 'Ow' from the airbag is going to be much worse."*

Before calling it a day, it's back to the killer tissue box. The final word is left to Buster, who is about to experience a head-on 120kmh collision with the Plymouth Fury. Adam: "Worse case scenario is we start up with the truck dragging the Fury and the Fury goes waaay off and smacks into something of value … like a crew member."

Jamie: *"This is actually really scary – 1800kgs of car rolling on its own at 80, 90, 100kmh is a scary thing."* One end of the cable is attached to the Fury, the line is then threaded through holes in the barriers and the other end tied to the pick-up.

Jamie floors the accelerator.

"Okay, this is the 120kmh test – the maximum speed!" – Jamie

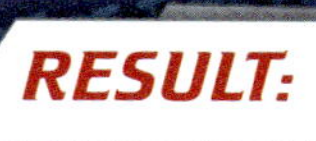

RESULT:

A TISSUE ISSUE

The Fury chews up tarmac and hits the barrier at more than 100kmh. Crash. Laughter. On impact, the Fury leaps up in the air and the tissue box's trajectory takes off from the back of the seat and smacks into the back of Buster's head. The speed at impact is 105kmh.

Amazingly, the box survived intact, the cardboard is not crumpled, there's not even a single torn tissue. There's not even a Buster-shaped head imprint on this. Grant plugs into the accelerometer implanted in Buster's head. Grant: "114.8Gs!"

That's very significant (remember an astronaut pulls only 8Gs during lift-off).

There are two smacks to Buster's head. One shows he turned and hit the dash, which was damaging, the second was the impact of the flying tissue box, not so harsh.

CONCLUSION

Adam: *"I think killer tissue box is definitively busted. Even if you had a decorative plastic or metal cover on this thing, it just doesn't have the mass to inflict lethal damage."*

Jamie: *"Absolutely."*

M = MASS (SLUGS
V = VELOCITY (FEET PER SECOND)
MAIN VARIABLE
T = TIME (MAXIMUM AMOUNT OF TIME FOR DEFLECTION)
man
Pascal = Newton/sq.meter
160 lbs
START
Crack

Rain, Rain, Go Away

Chapter 7

WALKING OR RUNNING IN THE RAIN: WHO GETS WETTER?

The Myth: ***You stay dryer if you run rather than walk through rain.***

Why are Adam and Jamie browsing in a sex shop? Reason: They are checking out latex body suits. This myth requires them to walk and run through rain and weigh the water in their respective overalls.

When it comes time to run the test, Adam and Jamie are going to have to make sure their perspiration doesn't mix with the rain so they'll have to wear something under their overalls. Something snug and form-fitting, which explains their visit to the land of latex.

Jamie asks Adam if he's ever worn latex before; Adam admits he hasn't. So Mr. S Leather store manager Richard Hunter explains to them the properties of latex – how it stretches, doesn't tear, its incredible strength and how you step in and out of your costume. Though a little unsure at first, Jamie tries on a body suit and seems quite taken.

"The results aren't what I expected at all."– Adam

He says, "This is a kinda cool because it completely closes up around you."

"How does it feel?" asks Adam.

To Adam's amusement, Jamie replies, "It feels kinda sexy!"

A CONTRARY OPINION:

The National Climactic Data Centre Test

Over the years, several studies have sought to answer the question of whether you should amble or gallop through the rain. Meteorologists Thomas Peterson and Trevor Wallace conducted their own experiment at the National Climactic Data Centre in North Carolina. They simply waited for a downpour and stepped outside.

Thomas Peterson: *"We measured an approximate course to try and get 100 metres, but we didn't feel it was crucial to get 100 metres because we weren't trying to see how much rain we absorbed in 100 metres but the relative difference between running and walking. Tom walked and Trevor ran, then they weighed their clothing.*

Trevor Wallace: *"According to our model we concluded that it's much better to run rather than walk. The runner was 40% less wet than the walker."*

The question is: *Who is right about walking or running in the rain? Is it Mythbusters or the National Climactic Data Centre?*

So they subsequently revisited that story with Scottie, Tory and Kari, who came up with the opposite result.

Says Peter Rees, "I don't know why we got that result in the first experiment. We did everything meticulously. That result still stands but obviously the real rain wins out."

The other interesting thing about this program is that if you look on the spine of your Mythbusters DVD, the logo features Jamie in the sex shop gear that he wore for this episode. Peter Rees: "Well, San Francisco being San Francisco – and sex shops being what they are – we did go through a period where we found them useful because there's a whole range of stuff you can get in sex shops that you never get anywhere else.

"We ended up getting a couple of fisting dildos for our shark punching rig for the Jaws special that we did *(one that could jab and one that could punch)* and they were quite expensive, $250 a dildo. And, although that doesn't get quite explained fully on the Discovery Channel, we are now in the possession of a robot that punches sharks with dildos.

"And those gimp suits Adam and Jamie were wearing were perfect. We could have got the helmets and the zipper across the mouth and the whole bit, but we wanted to hold back a bit for Discovery."

There are two conflicting ideas about who gets wetter:

1. A person walking through the rain will stay in the rain longer but the rain will be consistently falling on their head and shoulders, or;
2. Someone running through the rain gets wetter because they are picking up rain on their entire front.

THE SET-UP

Will they just wait for the next rainy day and go for a stroll? No, to test this properly the Mythbusters will need controlled conditions.

Jamie says, "We're going to have to be really careful about this. It seems like a simple thing, but there are actually a lot of factors to consider. You've got the velocity of the rain, you've got the speed that you're running and all these different things to factor in there. The hardest part will be to create very reliable consistent rain, something that's realistic. It's got to be the right velocity, the right drop-size, and a nice even spread over a big distance for us to get a good consistent run."

They decide to set-up the experiment in a 30-metre-long hangar which Adam and Jamie will first walk, then run through in a downpour of home-made rain. They will wear identical cotton overalls, which will be weighed after each test to see how much rain they soaked up. This means they'll have to manufacture some very precise precipitation.

At this early stage what are Adam and Jamie expectations?

Jamie says the walk will make you wetter, "because you spend more time in the rain".

Adam is not so sure: "I'm going to disagree with you just for the hell of it. I'm going to say the person who runs through the rain gets wetter."

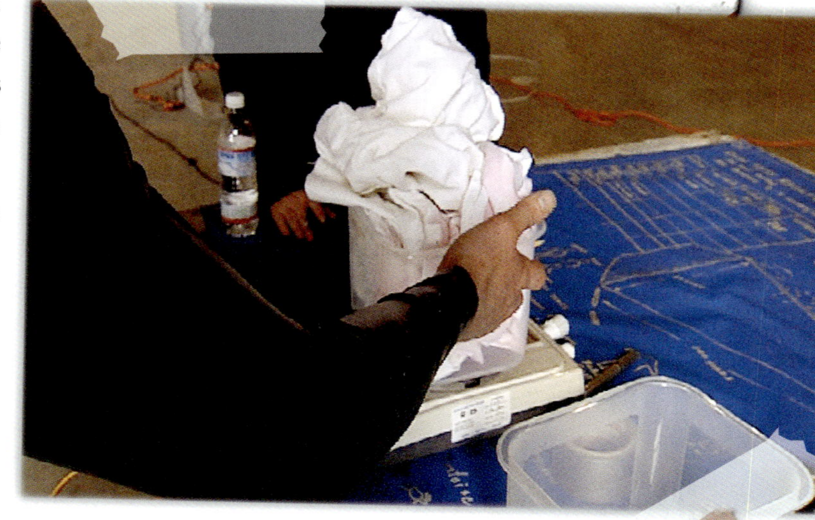

Having run (and walked) through the rain, weighing the overalls was the critical part of the process.

"Those gimp suits they were wearing were perfect. We could have got the helmets and the zipper across the mouth and the whole bit, but we wanted to hold back a bit for Discovery." – Peter Rees

The next step is to buy some piping for their home-made downpour. So they take a drive to a familiar plumbing store. When you've done special effects for as many sci-fi movies as Jamie and Adam you see irrigation equipment in a completely different light. Jamie knows his requirements precisely. He purchases 12 metres of 12mm pipe, and 45 metres of 50mm pipe. When he walks out, ironically, it's raining outside.

1. Pipes: The aim is to get about 20 metres of height so that the raindrops can achieve terminal velocity. Falling from 20 metres a drop of water will travel at maximum speed, which is around 6.7 metres per second. After that it cannot go any faster, no matter how far it plummets.

Jamie explains why he needs this construction rather than a simple garden hose. He says, "If we just go up there with a hose and squirt it down it's not like real rain, it might be going faster or slower."

The Alameda Naval Base is a big hangar, across the bay from San Francisco. It has quite a military and film history, and more significantly it has sufficient undercover space for the Mythbusters team to build a long enough run for the experiment, which requires 45 metres of pipe with sprinkler heads every 180cm.

Adam is pleased with the way things are going; there are plenty of tie-points and the pipes look really stable. He completes the rig and says, "So long as Jamie holds up on his end, we shouldn't be here too long."

No half measures: for this myth to get busted, they needed a controlled environment. Hence, the hanger.

2. Fake rain: In this experiment they're looking to duplicate average rainfall, which is between 5-7cms per hour. Jamie is working on getting water for the challenge – which is not as easy as you might think on a rainy day. He uses a fire hydrant and explains, "It's legal to tie in to these fire hydrants but we had to go through the Fire Department, who rented us a meter that allows us to keep track of our water usage."

Accessing water is one thing. Getting it to travel 20 metres straight up in the air is another. Jamie has calculated that the water will have to be pumped at 1 kg-force per square centimetre, which is no problem for this pump which can move H2O at a rate of 750 litres per minute at 3.5 kg-force per square cm.

3. The course: The third part of the experiment preparation is simple, they have to mark out a 30-metre course which they do with gaffer tape, marking out the start and the finishing line.

4. Fan plan: Finally, for added excitement, the Mythbusters want to add wind as a variable for some of the tests. For a while it looks like they can't agree on a fan plan – Jamie wants to get the fans as close to the course as they can. Adam differs, reasoning that consistency is more important than the highest wind speed. "Let's set it up my way first," he says, "Do you want to set it up your way first?"

"I don't care, actually Adam," says Jamie in a tone that suggests an argument has been avoided.

5. Gauges. Digital gauges are set out along the course. Adam is pleased with the equipment: "Oh, these are fabulous, it's almost as though whoever invented these didn't want to get wet when measuring the rain! I can't tell how big the drops are, so we're going to run it for five minutes."

After a preliminary test Jamie announces they're getting about 4.5cm per hour. "We want to goose up the throttle just a little bit and we should have the perfect amount of rain," he says.

And finally: Adam says, "Are you ready?"

"Ready as I'm gonna be," Jamie responds.

The time has come to see whether or not Jamie and Adam's rain-delivery system is just a pipe dream.

THE EXPERIMENT

Now the overalls are weighed. Each pair weighs 757 grams. Adam and Jamie wear them over their latex body suits.

It's time to walk the line. Their high speed camera is set to run at 1000 frames per second to capture every single droplet. Adam and Jamie will walk the course twice, once with wind, once without. Each take is timed, then the overalls are taken straight off and weighed. Both walks clock in at around 18 seconds.

- **Result:** 785 grams. The water content is 18 grams of water and both Jamie and Adam's overalls have soaked up the same amount of water.

Water, water, everywhere ...

But what happens when Adam and Jamie up the pace?

- **Result:** Adam's run yields 798 grams of water, which is significantly more. At 793 grams the difference isn't as great on Jamie's run. But the raw data still indicates you get wetter if you run than if you walk.

The difference using wind blown rain is minute – it only makes the overalls a few grams heavier. The overall result, however, remains the same.

"I'm going to disagree with you just for the hell of it. I'm going to say the person who runs through the rain gets wetter." – Adam

CONCLUSION

Adam: *"The results aren't what I expected at all, the numbers are really close, we're only talking about a matter of a dozen grams here and there. So when you average it all out I think it's going to show that you actually get wetter running – strangely. So what's the verdict?"*

Jamie: *"It's better to walk than run. It was very clear – over a 30-metre course with average rain delivery we got more than twice the rain per foot on the running than we did with walking.*

Adam: *"Is this one busted?"*

Jamie: *"This one is busted. The fact is that it's better to walk than to run in the rain."*

M= MASS

Natural Gas

COULD YOU BLOW UP YOUR OWN STOMACH?

The Myth: ***Six packs of Pop Rocks and six cans of soda is a lethal combination of gases that will make your stomach explode.***

In the annals of medicine in the last century that there are a half-a-dozen cases of people's stomach's exploding from an excess of CO2 – usually from something like baking soda.

"Do you remember Pop Rocks?" asks Adam.

"Yes," replies Jamie, "they were invented in 1956, the year of my birth."

Pop Rocks is a carbonated sugar candy exposed to a mixture of pressurised carbon dioxide gas; they are allowed to cool, trapping the pressurised gas inside. When put in the mouth the candy breaks and melts, releasing carbon dioxide from tiny 600psi bubbles which results in a popping sensation before swallowing.

"Have you had them before?" asks Adam.

"A long time ago, yes," says Jamie.

Adam explains, "The legend we are approaching now is about Little Mikey from the Life Cereals commercial. The commercial had the punch-line, 'Let's get to Mickey to try it' because Little Mikey hated everything, and whatever Mickey could eat, everyone could eat.

"Then Mikey became a teenager and allegedly drank six cans of soda and ate six pouches of Pop Rocks at a party. The two substances were said to have created some kind of chemical reaction in his stomach which exploded, killing him horribly. This urban myth is why Pop Rocks were taken off the market in the early eighties."

'POP' GOES THE SALES!

Vice President, Pop Rocks Inc., Fernado Arguis, explains company history:

"Originally the idea was to create an instant carbonated soda by General Foods. They discovered when you ate those nuggets there was a popping sensation in your mouth and in 1975 we decided to launch Pop Rocks.

"The difference between normal standard hard candy and Pop Rocks is that when the candy actually is still melting you carbon-diox it under certain conditions of pressures that I believe is 600 PSI. The myth appeared in 1979, four years after the candy was launched, and it had devastating consequences for the brand name. Sales dropped down to almost nil. The company tried to address the problem. They ran 145 full-page advertisements in magazines and wrote 50,000 letters to different high schools' headmasters explaining there was no problem."

Peter Rees explains, "The important thing about this story was – when I wrote the initial proposal for the first three episodes – *Pop Rocks & Cola* was one of the stand-out ones. I mean, it's a classic acid / alkaline school science experiment."

The story of the Pop Rocks Company is interesting, because the myth completely destroyed the company. Peter Rees shakes his head: "All I can say is that when we last rang, the guy that we interviewed no longer represents the company, so maybe they didn't want to revive those old myths."

This is a strange turn of events, because those who know the show will admit that not only did the *Mythbusters* make the product safe, but you felt you'd missed out on something if you'd never tried Pop Rocks.

"I totally agree with you," says Peter Rees. "But one of the problems in making *Mythbusters* is that the manufacturers of products that become famous don't want to know anything about it, positive or negative. It happens a lot, products don't want to be associated with anything positive or negative because some of these myths – even if they haven't been destructive to the company – can cause a lot of trouble. The cost of managing it is huge, whether it's good or bad for the company."

So, the Mythbusters would like to see how much pressure is generated by rapid infusion of six packs of pop rocks and soda. Second, they'd like to determine how much pressure a stomach can withstand.

"I've been thinking about this," says Adam, "and I think to imitate the pressures of a stomach the best course is to use a pig's stomach, which is actually physically very close to a human stomach (which can hold up to a litre, generally, of liquid)."

What follows is a series of humorous phone calls with Adam explaining to obviously astonished recipients that he's in the market for several pig's stomachs – "With as many of the attachments as possible" – by which he means the oesophagus and the intestines. "Hello," says Adam, "I'm trying to locate a pig's stomach. Shall we continue further?"

We can only imagine the person's expression on the other end of the line as Adam describes it as "like a very meaty swimming cap!"

It turns out that the Food & Drugs Administration have listed the oesophagus and bowel as hazardous substances. Jamie considers killing the pig, only for Adam to talk him out of it on the grounds of political correctness.

Jamie: *"It's better to kill the pig first and then give it to us?"*

Adam: *"Absolutely, most people are much more comfortable with that arrangement."*

So they go for a drive to a butcher who sells them four stomachs, complete with oesophagus and intestines.

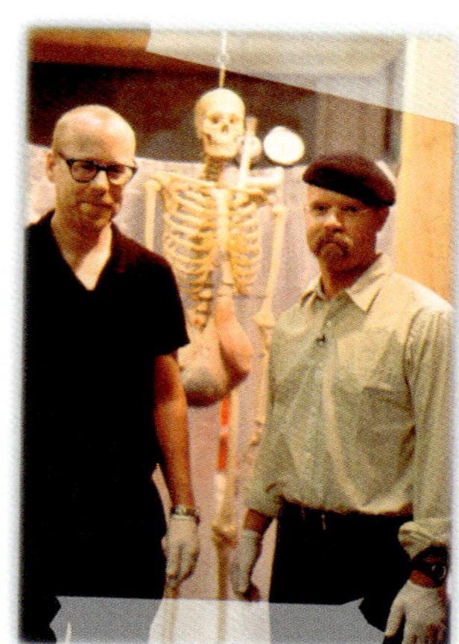

Arriving back at the workshop, Jamie is impressed that the pig stomachs come with specific instructions.

"This nightmare is becoming a reality," adds Adam. That's one important mission accomplished.

1. Inserting the stomach contents

Now wearing rubber gloves, Jamie probes with his fingers trying to determine which is the input and the output, after which the Mythbusters devise a plan of action: "To get these Pop Rocks in there we're going to have to get them wet.

I mean, they're not going to pour. They're going to stick along the side, they're going to immediately clog the hole." He gets the idea to insert them by wadding them up in little packets like tea-bags and dropping them into the stomach. But the packets are too bulky to be pushed through the intestinal tubes. So they agree on using a big syringe.

2. Hanging the stomach

The next part of the set-up requires hanging the stomach. Even as they raise the problem, the workshop skeleton is staring at them in the background, "What about mounting it in our skeleton?"

"Sure," Jamie replies. "We'll get it dribbling out of the skeleton and if you want a dramatic effect, that'll do it!"

Adam: *"And of course the inlet tube will have to come up and out through the mouth of the skeleton – and you know where the outlet tube's going!"*

Jamie: *"Where is the outlet tube going?"*

Adam: *"Don't make me say it."*

"Don't make me say it." – Adam

3. The gastric juices

Before the experiment can begin the Mythbusters will need to fill the stomach with gastric juice with a PH of 2 using hydrochloric acid.

According to gastroenterologist, Dr Stanley Benjamin: "Gastric juice is secreted actually, we believe it's a protective mechanism and not designed to aid digestion. Hydrochloric acid along with pepsin is designed to kill things, so it's a kind of protection when we were back there fighting hyenas for the bones."

It might not aid in digestion but we think a quart of hydrochloric acid will be critical to busting this myth. Let the fun begin!

THE EXPERIMENT

There are two parts to the experiment:

1. To see how much pressure is generated by filling the stomach with six pouches of Pop Rocks and six cans of soda drink, and;

2. If that doesn't cause the stomach to rupture, what does?

This is a messy – no, a disgusting – job, a pig's stomach, suspended in a skeleton. Most unusual TV viewing.

1. Pop Rocks and Cola.

The orifices are now scissor-clamped, the Pop Rocks inserted and it is agreed that, while Jamie pours the liquids, Adam will open the cans and hand them to him. After two cans the stomach begins to pop. Four has Adam yelling, "Oh my god, he's going for another one. Mikey! Dude, you are an animal! You can't drink those last two cans of soda! Dude, your stomach's going to burst!" Jamie is calmer, "Slow down, he's not drinking that fast. You can hear those Pop Rocks going – that's pretty impressive!" Then the stomach begins to burp, after which it begins to distend. "That's huge," says Jamie. "That's like three times the size of what it was."

Working with all that gas, a little time outside was necessary.

But, despite six cans of soft drink, six pouches of Pop Rocks and a litre of acid, there is insufficient pressure. Gastroenterologist Stanley Benjamin explains, "By its nature the human stomach is a capacitance organ, that means it's designed to stretch. Receptive relaxation is the concept in human physiology so it will relax and stretch and stretch up to some limit." Even with time for digestion the Pop Rocks and soda just don't produce enough gas to burst.

2. Bicarbonate of soda. After a quick reset, the Mythbusters are going to try and see if they can make a stomach burst. They're going to test how the baking soda compares to Pop Rocks in terms of expansion. First they need a new stomach, secondly they need to clamp it, third they have to re-create the natural stomach acids. Then Adam pours out three tablespoons of baking soda.

Jamie: *"That's an awful lot of sodium bicarb."*

Adam: *"Well, I've taken two with heartburn."*

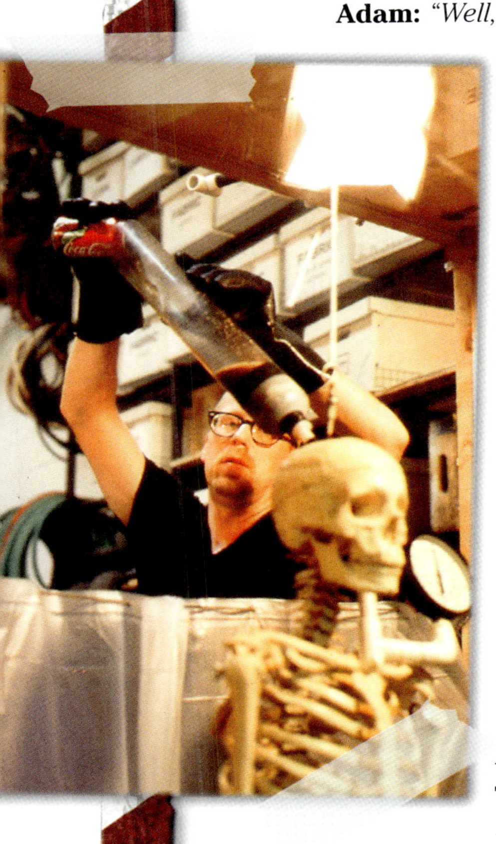

Adam takes care to introduce the right amount of soda ... six cans in all.

Unlike Pop Rocks, the baking soda and acid reacts. The result is carbon dioxide gas. Lots of it. Next they need the stomach to ingest six cans of soda. "Okay, one down," says Adam, still pouring. "Three cans down ... four. C'mon, drink a little faster for me. Five ... six! Omigod he can't hold it any more. Clamp it!" They check the pressure which – although much higher than before, is still hardly giving a reading, it's reading about half a PSI.

3. Pushing the limits. Now, the Mythbusters decide to push the envelope. Additional teaspoons of bicarbonate of soda will be added until the stomach finally bursts. It only takes two more.

Says Jamie, "Well that did cause a reaction. If you have a stomach that's greatly distended by a large meal then you add on top of that gas, in the form of sodium bicarbonate, to try and relieve acid indigestion, you can in those very unusual situations see gastric rupture. But no-one should have the impression that this is common by any stretch of the imagination." Adam sums up the effect: "That turned out to be a little too much. That's what we call a helluva Saturday night! With the addition of more stomach acid and the bicarbonate of soda ... It ripped a hole right in the back, I felt the pressure as I was plunging it, it wanted to push the plunger back out again. I don't know if you were watching what the meter said but it might have spiked. In a body something like 2-3 PSI is a huge amount of pressure. I can't imagine a human consuming that much crap ... maybe I can, actually."

RESULT:

NATURAL GAS

MYTH:

CONCLUSION:

The human stomach clearly can rupture, there's no question about that. There are probably lots of folks out there who have for years and years been using baking soda, and in the amounts prescribed to reduce acid there is very little gas formed. It usually stays in solution, it doesn't cause very much by way of distension if you stay within the limits that are recommended.

So does this myth contain any elements of truth?

Jamie: *"The sodium bicarb seems to me to be the main thing. Somebody that's just eating candy, I don't see it happening."*

Adam: *"Another one bites the dust."*

FOOTNOTE: *Mythbusters tried to contact John Gilchrist, the actor who played Little Mikey. Clearly he did not die. He is listed on the internet as currently working as an advertising-account manager for a New York radio station. He never returned their call.*

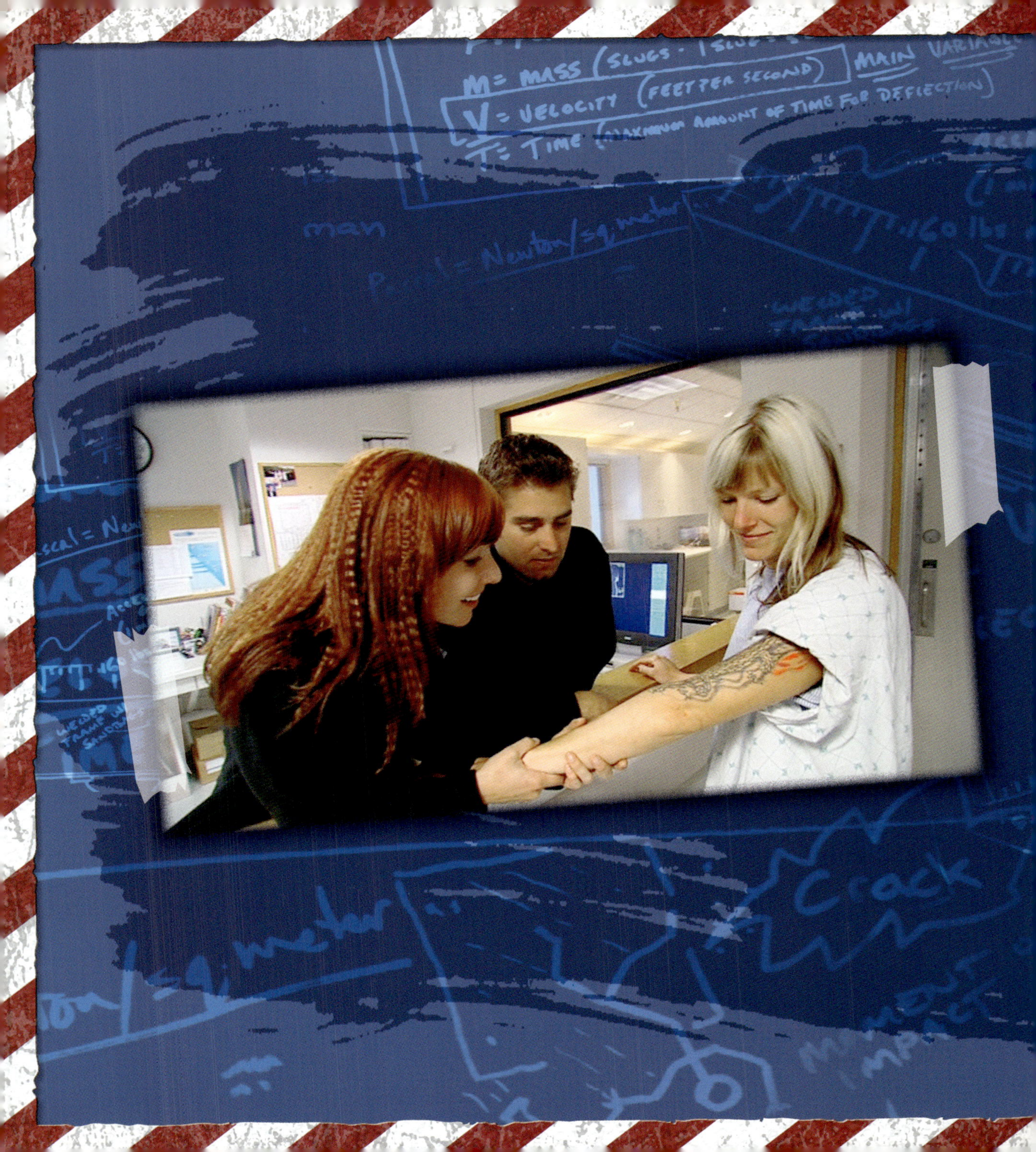
M= MASS (SLUGS
V = VELOCITY (FEET PER SECOND)
MAIN
T = TIME (MAXIMUM AMOUNT OF TIME FOR DEFLECTION)
man
Newton/sq meter
Crack
sq meter
MOMENT
IMPACT

Moving Pictures

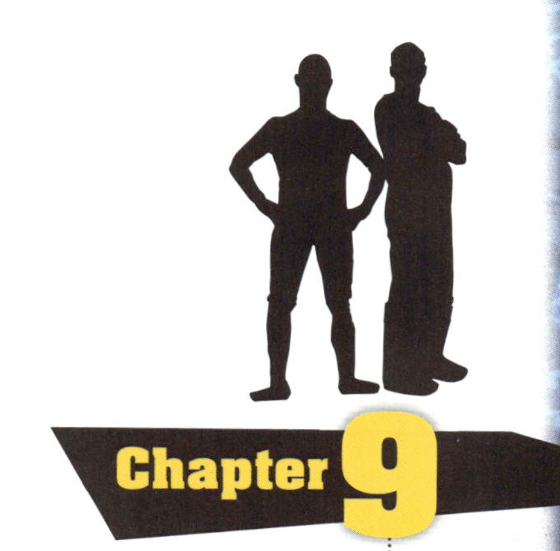

Chapter 9

COULD YOUR TATTOO EXPLODE?

The Myth: ***That a decorative tattoo explodes when undergoing a medical scan in a Magnetic Resonance Imager (MRI).***

Magnetic Resonance Imaging (MRI) is one of our most powerful tools for clinical diagnoses. Neurologists and radiologists use Magnetic Resonance Images to track the progress of degenerative disorders such as multiple sclerosis.

In this instance the Mythbusters needed to get on with official bodies and bureaucracies – often it's the Fire Brigade, but in this instance it's the Abdominal Imaging Division at the University of California.

Peter Rees explains, "This was one instance where we didn't have any trouble convincing them that it was a good thing to do. We were taking over valuable diagnostic time from real patients. But the one thing they didn't show was that we tattooed a pig with black powder and then set it off, to make the tattoo explode out of the surface of the pig. It worked, but it didn't get to air, so it must have been a bit gruesome."

"Could your tattoos explode?" is a myth with some truth to it. Apparently people get burns because tattoos have iron in them, and the iron does react aggressively with both the electro-magnetic field in the MRI machines and with the radio frequencies that the machine puts out.

Despite the intellectual rationalisations for the episode, Peter Rees cuts to the quick and says, "Let's face it, when you get to tattoo a side of pork and put it in an MRI machine – are you going to say no?"

Being a heavily tattooed lady, Scottie got quite excited over this episode, and was getting a whole dragon put on her arm at the time. Furthermore, as a metal worker, she had a high risk that she actually had tiny chips of metal in different parts of her body, particularly in her eyes.

And going into an MRI unit with any metal in your eyes isn't good. These magnetic resonators pull 30,000 times the earth's magnetic force. Eyes can explode, so she had legitimate reasons to be excited about doing that experiment as well.

OUTLINE

Tattoos which are more than 20 years old contain traces of metal. An MRI unit is basically a powerful magnet with a force at least 30,000 times greater than the earth's magnetic field. When metal fragments in the tattoo meet the MRI magnet – so the story goes – the result can be explosive.

"The iron oxide got sucked into the magnet! Thank goodness we taped the top otherwise I don't think they'd let us come back."
– Kari Byron

Fergus Coakley, Chief of Abdominal Imaging at University of California, San Francisco (UCSF), says, "One or two patients out of the hundreds who have tattoos and then are scanned do have a reaction, but the reactions that have been described are all minor, consisting of slight discomfort and redness around the tattoo. But nobody has described exploding tattoos."

Could it be possible that the magnetic field can induce some sort of current in the pigment material in the tattoo?

Coakley: *"It's possible, but it seems unlikely, because usually to get induced currents you need lengths of wire or loops. So it's hard to see how these tiny particles could have currents induced in them, though nobody could say it's not happening for sure."*

THE EXPERIMENT #1

Mythbuilder from the Build Team, Scottie Chapman, is a well-tattooed MythBuster, and she is prepared to put her body on the line for the show. After making appropriate arrangements with the UCSF hospital staff, she changes into a nightgown and enters the MRI cylinder. At the first sign of trouble the scan will be stopped.

An MRI scan targets the hydrogen atoms which line up with a strong magnetic field and are then bombarded by radio waves. The radio waves are then absorbed by the atoms and the resulting signal sent to a computer, which turns the data into a 2D image.

Result:

1. There is no distortion in the tattoo image.
2. Scottie experiences no pain, no discomfort – and there is no explosion.

But of course the tattoo myth isn't busted yet.

Scottie *(middle)* was the perfect subject for our MRI.

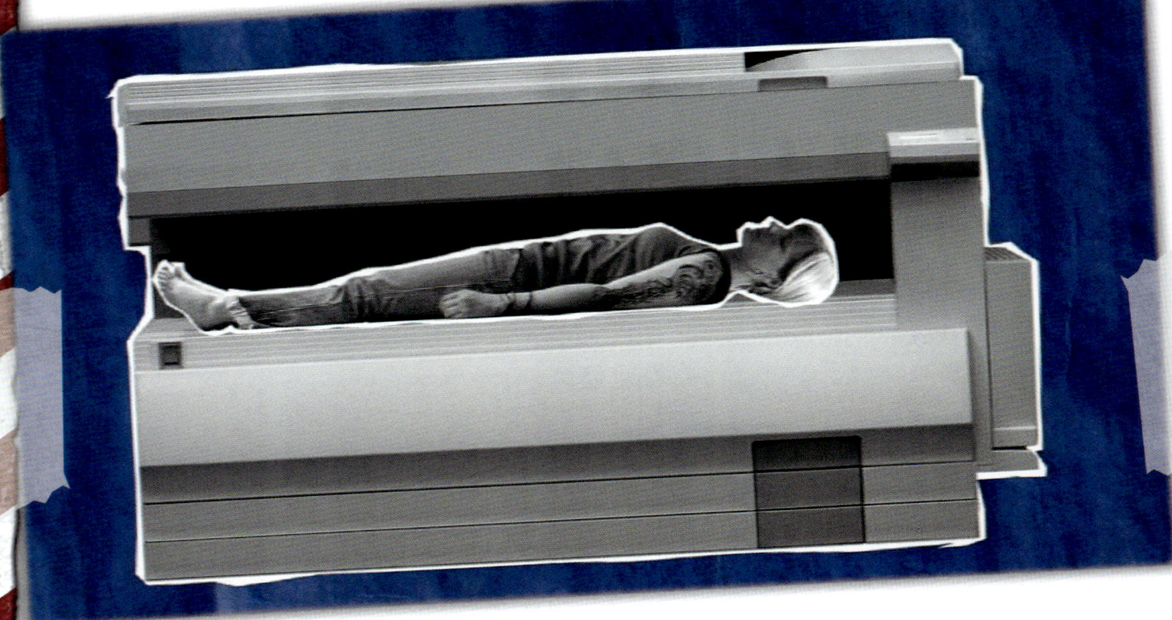

So, you've got tattoos ... and someone suggests that to test whether they will explode in an MRI, you should be the guinea pig. At what point was this a good idea, Scottie?

Did you know?

Why do statistics show that more women die using mobile phones at petrol pumps than men? *The reason is probably because of static electricity and because women talk on the phone when pumping petrol more than men. It is not because the mobile phone causes a spark which ignites the petrol vapours and causes an explosion.*

THE EXPERIMENT #2

To bust this myth they are going to have to individually test all the compounds that go into all the pigments and find out what could cause any sort of irritation or explosion. The theory is that the MRI magnetic field causes a reaction with the metallic components that are used in tattoo pigments, which are a pot pouri of chemical compounds including iron oxide, mercury, barium, zinc sulphate and titanium dioxide.

The Build Team are now going to expose some of these materials to a strong electro-magnetic field, which will enable them to see if the pigments line up with the magnetic field. They will then need an electrostatic voltmeter to check whether a current is being induced; and then also check any temperature difference if the pigments do create some sort of excitation in the scanner.

Result:

One by one they test iron oxide, mercury, barium, zinc sulphate and titanium dioxide. But only one – iron oxide – reacts, and it does so strongly. The iron oxide particles actually line up under the influence of the magnetic field. Bringing the iron oxide to the MRI machine, the iron oxide takes on a life of its own. There is a mini-explosion.

Says Kari, "The iron oxide got sucked into the magnet! Thank goodness we taped the top otherwise I don't think they'd let us come back."

The voltmeter exposed iron oxide for its strong personality.

Says Tory, "It looks like iron oxide is definitely the culprit here."

"The one thing we did show was that tattoos can interfere with MRI systems and make diagnosis difficult. And what's more there can be ramifications for people: if you do feel any sensation when you're in one of these machines you should speak up and not be afraid to say something. But if you didn't, it wouldn't kill you." – Peter Rees

THE EXPERIMENT #3

The Mythbusters still want to make a tattoo explode in an MRI unit. As Scottie won't be the guinea pig, they decide to use the cuts from real pig, which Tory, Grant and Scottie purchase from a local butcher – with the skin left on.

Scottie also introduces the team to Mattie the tattoo artist, who is responsible for the new ink on her arm. Mattie has heard the MRI myth but he doesn't believe it. He says, "I've never heard anybody say, 'I went and got an MRI and my tattoo exploded'. I think that would be pretty big news."

The pork is cut into portions. Mattie draws the word 'Mythbusters' on each portion using first a special brew full of iron oxide, the compound that reacted to the electro-magnet, and second a regular tattooing pigment.

The pork bellies are primed for scanning. If a tattoo is going to explode, it's going to be this one.

All three experiments put paid to yet another urban myth.

Result:

There's a glitch on the image – the heavy iron oxide content seems to be having an effect. The tattoo is definitely causing an interference on the hospital scan. But it's not explosive. And no pigment has been pulled to the surface after blotting.

Next is the traditional iron oxide you'd find in a normal tattoo parlour.

The iron oxide in the regular ink also interferes with the scan but the effect is much smaller. So, ultimately, Scottie, Tory and Cary are left with chunks of unexploded tattooed pork.

"I've never heard anybody say, 'I went and got an MRI and my tattoo exploded'. I think that would be pretty big news." – Mattie (The Tattoo Artist)

CONCLUSION

Busted or not? Says Adam, "Tell us what you came up with Scottie."

Scottie replies, "Clearly iron oxide is the culprit here. Iron oxide is very common in tattoo inks, and it definitely reacts to a magnetic field, but at most it's going to cause heating, it's not going to explode."

Jamie asks for confirmation, "So you're telling us there's no exploding tattoo?"

Scottie, "No Jamie, I know it disappoints you but I'm a little relieved about that. I'm saying myth busted."

There was more chance of producing a plate full of crackling than confirming this myth ... has someone been telling pork pies?

V = VELOCITY (FEET PER SECOND)
MAIN VARIABLE

Is Air a Con?

Chapter 10

IS AIR CONDITIONING MORE FUEL EFFICIENT THAN HAVING THE WINDOWS DOWN?

The Myth: ***It's more fuel efficient to run the air conditioner than to have the windows open.***

Ever since the first auto air conditioners arrived in the 1940s drivers have asked the question – is it more efficient to run with your air conditioning on full blast and your windows up than it is to drive with your windows down and the AirCon on?

With world oil prices on the rise, everyone's looking for ways to save pennies at the pump. That's what makes it a perfect test for the Mythbusters.

Jamie, *"AC uses an awful lot of power."*

Adam: *"A car with the windows open is creating an awful lot of drag."*

The Mythbusters are about to run two tests to find out once and for all.

This air-conditioning story is actually doing a community service. Everybody wants to know whether it's more fuel efficient with air con on or windows down?

"Any story about gasoline and mileage grabs people's interest," says Peter Rees. "In America, the two types of stories we get the most hate mail about are guns and cars, and that's because all Americans are experts on both of those subjects. Air con is such a great story because everyone thinks about that. I don't know how many times I've been driving in traffic in a Sydney summer thinking 'I wish I had air-con – but it drains too much power, blah blah blah ...'"

And the Mythbusters came up with two results ... which was very controversial.

It's all about controversy ...

"Yes, it was controversial, because we got the interpretation wrong and everyone knows it, and everyone's got an opinion," Rees continues. "So we revisited it. In America, the myth is that it's better to leave the tailgate down on utes because there's less drag because the air can flow free and through the back of your vehicle. And rather than doing a five-gallon test this time we did a 700-kilometre test run which definitely showed it's better to leave the tailgate up, a result supported by both Toyota and all the major car manufacturers. It's totally counter-intuitive, there's no way!

"When we first saw this result we said, 'We've revisited this bloody petrol thing and we've got it wrong again – how can this possibly be?' Personally, that's the territory that I would always want the show to be in and we're probably in that territory maybe 60% of the time, and that's where it's really interesting.

"Obviously when there's something you can't explain you know viewers are going to crucify you for it, irrespective of whether you're right or not. And the re-visits allow us to go back and add to stories."

"This test will not be fun – I'll have to pee and probably have to poo before this thing is done. I'll hold it in, the producer will be taunting me on the radio the whole time, the wheels are going to be screeching." – Adam.

THE SET-UP

One problem with this episode was that it was originally intended that Adam and Jamie drive non-stop for eight hours until the petrol ran out. For any number of reasons – sanitary perhaps? – the test was run over a shorter course, using only 19 litres (five gallons) of petrol.

"You're looking at a recipe for disaster when you try to speed it up and go into half-measures, because when we reduced the fuel in the tanks to 5 gallons the viewers can question, 'Did they both have the same five gallons?'" Rees says. "And it's always better to do it, even though you question it at the outset and go, 'Eight hours in a car?' Adam is very generous, Jamie not so generous, but because neither of them are scientific, we have to make them actually be meticulous and understand you can't change the perimeters between experiments.

"Once you've set up that eight hours, you've got to stay with that eight hours, otherwise all your data doesn't mean anything. If anything, that's the take-home message from the whole show: we're not trying to say we are scientific, all we're trying to say is a logical thought process and there are certain rules that you can use. Watch us apply them or not apply them as the case may be. We're happy to suffer for it and we're happy to go back and do a re-visit if people think it's necessary."

THE SPEED THRESHOLD

"We got the facts wrong because there's actually a threshold there for speed, it's about 80kmh. Above that, the loss of energy through the drag on your windows far exceeds the sap on your energy that you get from your air conditioning. Below that it's better to have the air conditioning off and the windows rolled down for the vehicle." – **Peter Rees.**

First Test: At the Altamont Raceway an auto-computer will test fuel consumption by monitoring the rate of air flow through the engine of two Ford SUV (Sport Utility Vehicle) vehicles.

They are going to drive 15 laps at 90kmh with a computer hooked up to the car.

- five laps with the windows up;
- five laps with the windows down, and;
- five laps with the windows up and with the AC on full.

They will see if the computer shows up a significant difference in the mileage of the car under those conditions.

Result:

- With the windows up and AC on it was just on 5km per litre.
- The AC off it didn't make much difference.
- With the windows down it was 4.7 km per litre, the drag of the wind made the difference.

That seems pretty significant over a full tank of petrol.

CONCLUSION: Clearly it's more efficient to drive with the windows up and the air conditioning on than with the windows down. That's what the computer said. But there's still a question mark because those figures are based on a computer model that assumes ideal conditions.

Second Test: *(The 'Average Joe Test')*

Adam and Jamie are going to run two identical SUVs, with two full tanks and two identical loads. Jamie's grey SUV will have the AC on full, Adam's vehicle has the AC off and the windows down. The logic is simple: the vehicle that uses the most petrol will stop first. First the guys empty the SUV's tanks and put exactly 19 litres of petrol into each one. And away they go, sticking to a speed of 72kmh because of safety concerns.

Result:

Jamie's AC-cooled SUV grinds to a halt. Thirty laps later Adam's naturally-ventilated vehicle finally runs out of juice. Says Adam, "That totally confounded my expectations."

Jamie: *"There is no other way of interpreting it, we've gone over all the facts of what went on here, and I ran out first by running the AC."*

Adam: *"Yes ... 15 miles ago, that's 30 laps that I did that you didn't do."*

Jamie: *"That's a significant difference in consumption!"*

RESULT:

IS AIR A CON?

MYTH: BUSTED

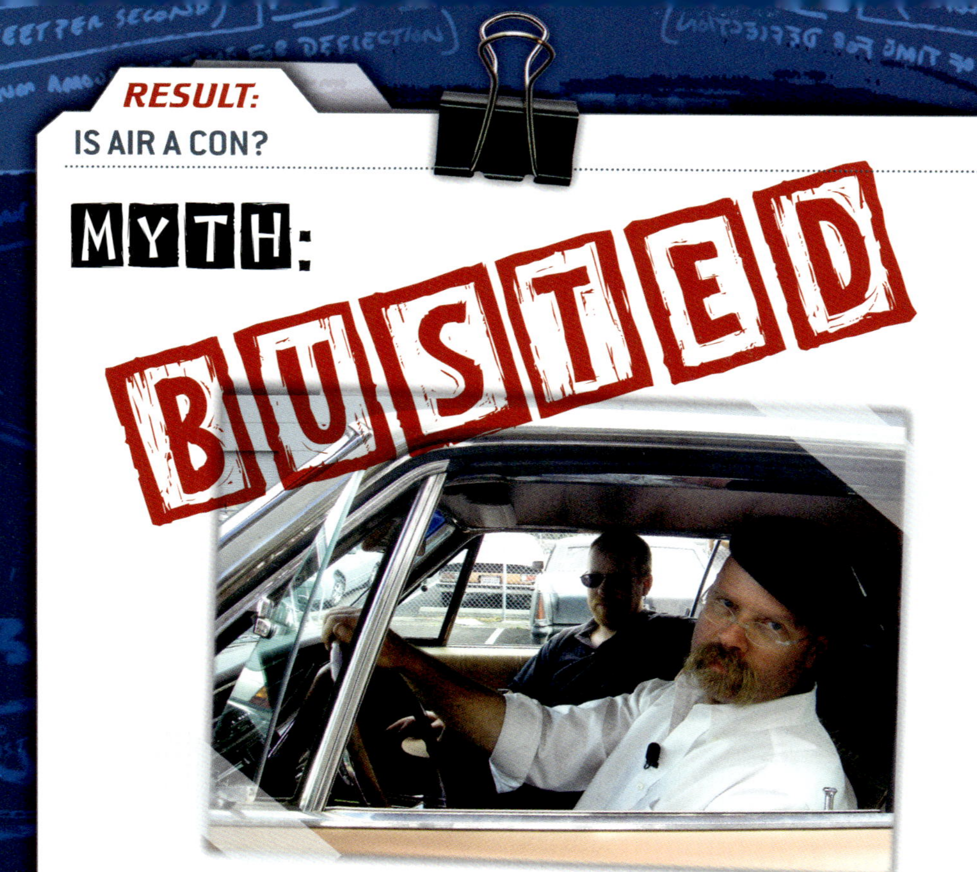

CONCLUSION

The computer showed it was actually better to run with the AC on, not by much but it was better.

Adam: *"I've got a little bit of a problem with that, which is the computer is actually measuring the fuel consumption of the engine, and basing its assessment on a computerised model of ideal circumstances. However, we definitively showed that running with the windows down was a lot more fuel-efficient by almost 15%."*

Jamie: *"There was a pretty significant difference."*

Adam: *"So AC being more fuel efficient? – I'd say busted."*

Jamie: *"Busted, yes."*

M= MASS
V= VELOCITY
(FEET PER SECOND)
MAIN
DEFLECTION
160 lbs MASS
WELDED FRAME W/ SANDBAGS

It's for you!

WHEN YOUR PHONE DELIVERS SHOCKING NEWS

The Myth: ***That a lighting strike on your house can travel down your home's wiring and kill you when you're on the phone or in the shower.***

Lightning fatality statistics bear out the dangers of electrical storms. On average about 100 Americans are killed by lightning each year, with 2% of those people talking on the phone at the time.

To test the dangers of lightning bolts on a house, Adam and Jamie have set up a little playhouse on wheels at the lightning strike testing facility at the Pacific Gas & Electric, which is a huge dome containing its own power generators.

This facility can generate up to 700,000 volts at 500,000 watts, but by using half that the *Mythbusters* will get more amps, which is essential for this test because it's the flowing current that does the damage.

Strangely, when the lightning strikes the phone with such force, another question might arise: are the computer and TV okay?

"The computer and the TV were okay, but they wouldn't be under normal circumstances – that's a glitch that wasn't explained fully in the story," says Peter Rees. "The reason they didn't explode was because you cannot have anything in the experimental area that has a chord leading away to another source of power, because if a computer was blown up, which it would have, it would have probably transmitted the power through that line and blown out their entire system.

"As it was, we blew up two or three of their static voltmeters, which they weren't very happy about, so they wouldn't let us plug in the TV and computer during the experiment. We're happy to try and do anything but sometimes there's some stuff that even we can't do and that is one of them.

"The difficulty with that experiment is that replicating the lightning is extremely difficult. The facility that we used was P&G which is a major energy supplier for the entire west coast of the United States. Even that facility can't generate simultaneously the same amperage and voltage as a lightning strike. They can either generate the voltage in one facility (and the voltage is the thing that gives the spark). Or they can generate the amperage in another facility (that's the thing that makes things blow up). So there's not really any facility in the world where you can replicate both of those things."

Adam likes nothing more than playing with big, powerful toys that generate a serious bang.

THE SET-UP

First, the playhouse is cut in half to enable the team to see what's going on inside. Inside, the Mythbusters have added a couple of common appliances for visual realism, but connection to mains power during a test would be dangerous, so they're left off. The playhouse has wiring, plumbing, working shower, TV, computer and wall switches.

Stephen the electrician gets Adam's wiring to work. When asked about melted phones, he admits he has never seen one.

Next they need a volunteer, and since everyone is keen to stay alive, Chip the ballistics gel dummy will be taking hits for the team. The gel has similar electrical resistance to human flesh and Chip is grounded to the earth like a person standing up.

Chip is set up with a phone strapped to his ear. For dramatic effect, Adam and Jamie have also taped some gunpowder into the mouthpiece. And finally they insert their Mythbusters-patented Hot Paddle Current Meter, where they set up two copper paddles right where the heart would be.

If they read more than 6 milliamps of current across these lines, they know they've got a dead guy.

"All TV producers, writers and directors go through a period where they want to blow up a television. It's in every music video, and it was my turn. I've done that now - I don't have to do it again."
– Peter Rees

Jamie explains the electrics. "The electrode will give us about 200,000 volts. That's enough to make it jump across the two feet through these wires, down into the electrical box or through the phone lines."

Adam: "We would be able to see the effect on a television or on a computer is if you left it plugged in during a thunderstorm, but unfortunately we won't be able to have them running during a test."

Jamie: "I think what's going to happen when we zap the house is we're going to see this nice long arc jumping from the electrode into the house. We may see a few little things sizzling and burning."

The Mythbusters have done their best to duplicate in several different configurations what kind of wiring a lightning bolt is likely to encounter but it's fairly unpredictable stuff; even the guys here aren't exactly sure what will happen. The only thing the experts can be sure of in this test is they don't want to be standing anywhere near it when it happens.

Everyone in the team is barricaded a safe distance from the structure, where they must remain during the test.

THE EXPERIMENT

With the electrode in this position, when the charge reaches between 2-300,000 volts it will zap across to the house. It's a waiting game as the charge builds. Here's how this will work – electricity can jump through air, and every 10,000 volts adds about one centimetre to the distance.

1. **300,000 Volts. Zap!** 300,000 volts straight into the housing wiring (standard to most homes) generates an electro-magnetic pulse strong enough to throw the remote camera out of focus. But there was no smoke, no sparks from the phone and no current to Chip as the charge travelled out of the ground. Incredibly, the phone call is uninterrupted. Jamie concludes, "Everything seems just like we left it. No problem, no reaction whatsoever. I don't think he got any current," but remember this shock is much smaller than real lightning.

Mini Myths

Dead body in a car: *The myth goes that if you leave a dead body in a prestige car for two months, you won't be able to sell that car at any price because of the smell, which cannot be removed. Outcome: with extreme cleaning you can get the smell out, but a decomposing body ruins seats, paneling, electronics – everything. This is more likely to be the reason the car won't sell.*

Setting up house was fun, but blowing it to smithereens was even more exciting.

2. The worst possible scenario. There's no facility in the world that can generate both the voltage and the amperage of an actual lightning bolt, so Adam cuts some of the grounds to the electrical wire system in an attempt to guide the electricity over to Chip. So, now minus the ground wire, can that thin telephone cord carry a fatal charge? Zap! The lightning arced into the mouthpiece and into Chip's head, setting off Jamie's charge. It tripped the fuse of the meter at 40mA. It arced through the mouthpiece and into the body. Jamie is convinced: "That's it right there as far as I'm concerned. This is a standard phone. It arced through the mouthpiece of the phone, that's a lot of current to go through gel. As 6 milliamps across Chip's heart paddles would indicate a fatal shock, therefore more than 40 milliamps means 'That's a dead guy'. So it seems it's myth confirmed – even with a modern phone and modern wiring you can be struck by lightning."

"Stay away from things that can conduct lightning to you – appliances can do it, water can do it, your plumbing can do it, stay away from those things." – Adam

3. An older style phone. As a further test they try an older phone and an older-style fuse box. Zap at 200,000 volts. This time the results are less dramatic, as Jamie's gunpowder didn't ignite, but Chips was definitely hit, and the meter blew again. Adam sums up: "We've definitely still got continuity through the body and into the ground." "I'm satisfied," says Jamie, "Phones will do it." For extra clarity, Adam adds, "Phones and lightning – bad, bad, bad!'" Ironically, when powered up, the TV and computer worked fine.

While it looks like fun on air, the Mythbusters take extreme caution playing with things like lightning.

4. The Shower. The Mythbusters have grounded Chip with an electrical wire to represent the equivalent to standing on a drain, which is also grounded. The electrical wiring is running next to some of the shower. Says Adam, "We're hoping for a short between the electrical wire and the plumbing." Zap – at almost 250kv. Now the guys learn that fluid streams aren't great conductors – the liquid breaks up so there's no direct path, but at these voltages … who knows? "Did you see that?" Adam exclaims excitedly. Jamie replies, "Yes I did, I'm not sure what I saw but I saw it." There were big electrical explosions in the shower, they even managed to start a fire in the shower. They put it out and Chip is saved from the fire, but was he already toast? The meter is inconclusive again, but the visual evidence suggests you wouldn't want to be in the shower during a storm. Says Adam, "I don't care what the meter says, we weren't even remotely approaching the actual strength of real lightning, and nobody can. And yet we're getting these massive arcs dancing across this guy's body." Jamie adds, "Arcs the size of boa constrictors! So you kinda get the hint that there's going to be a problem there somewhere."

RESULT:

IT'S FOR YOU!

MYTH: CONFIRMED

CONCLUSION

Adam: *"I'm looking at phone confirmed, shower plausible."*

Jamie: *"I'd say that's a safe bet. That amount of voltage and power can go anywhere, and so if you have a lightning storm you want to minimise the possibility of exposure to the volts."*

Adam: *"Stay away from things that can conduct lightning to you – appliances can do it, water can do it, your plumbing can do it, stay away from those things."*

FAN-O-DEATH
M= MASS
(FEET PER SECOND)
MAIN

Short Back and Sides

CAN A CEILING FAN CHOP YOUR HEAD OFF?

The Myth: ***That a domestic fan can cut off your head.***

What does it take to sever a human head from its body? Never mind the guillotine, could you do it to yourself? Might you have all the equipment you need in your home right now?

One of the things this episode suggests is that it is more difficult to chop off a head than you'd reckon. Nevertheless, every Aussie kid will identify with Peter Rees' childhood memory.

"It's really an Australian story," he says. "Every time I went on holidays to Nambucca Heads, the Sunshine Coast or the Gold Coast my mother was always saying, 'Don't jump on the bed, you'll take your head off with the ceiling fan!' And it's all bullshit, the fan doesn't have the velocity, it'll stop if you grab it."

Peter recalls the most frightening thing about the entire story was when Adam grabbed the fan. That's right, out of the blue, when no-one had tested the outcome, Adam did that. "He's done that kind of thing before," says Peter. "He did it with a

"I can't believe how hard it is to chop a head off." – Scottie Chapman

washing machine. And my mother always said, 'Never put your hand in a washing machine when it's running otherwise you'll get dragged in'. Adam, who understands the mechanics of it all, realises that the thing has a clutch; same with a ceiling fan."

As for chopping off someone's head … although this episode makes it seem difficult, it is too easy. Forensic scientists do their tests on various animals that best match the human equivalent of whatever they're testing and, for a human decapitation test, it's a sheep.

Says Peter Rees: "We actually got a $2000 samurai sword and we thought, 'Let's just use this for the purpose for which it's intended' and Tory had a go, and I can tell you it is frightening how easy it is to chop someone's head off with a sword.

"Fan blades are very flimsy, they don't have a lot of mass. And to actually get your head in line and your neck perfectly in position for where a blade would cut is next to impossible. What's going to happen is exactly what happened as you saw – you come up from below, it bounces off your forehead and doesn't cause much injury.

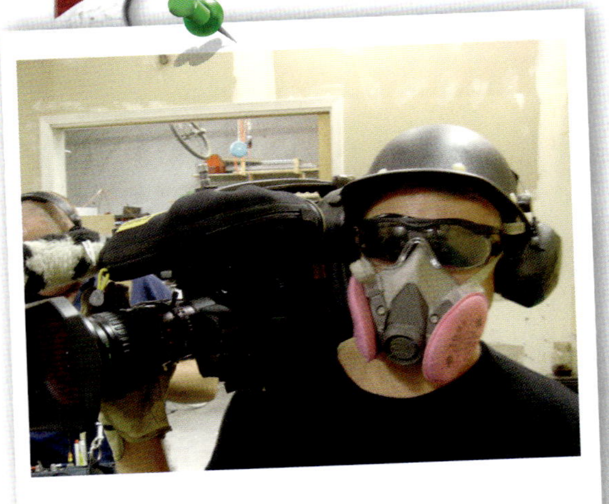

Can't be too careful …

"If you're coming from the side you're going to get this boom-boom-boom on your shoulders and you'll bend the blades before it gets a chance to get anywhere near your neck.

"Whereas the part of the sword that's cutting is a third of the way down the blade, it's not the tip, so that's why it is simple to decapitate someone.

"But with a ceiling fan it's completely different no matter how much you amp it up – as we did with the lawnmower through the roof."

The myth is usually told in two different versions:

1. In one, a husband and wife are getting frisky and he's all dressed up in a costume, and he leap towards her, jumps too high and gets beheaded by the bedroom fan.
2. In another one, kids are jumping up and down on a bed and one of them jumps too high.

THE SET-UP

Two myths, two rigs. The myth of the jumping kid calls for a rig to a raise a head into a fan from below. For the myth of lover's leap, they'll need a rig to carry a head into the ceiling fan at neck level, which begs another question. "Where are we going to get a head?" asks Tory. Ballistics gel, of course.

1. **The human head.** "We're going to have to duplicate a head," says Adam. "We've often done it with ballistics gelatin but in this case we're going to have to put a real spine in it. Obviously we won't be able to get a human spine – it may be gross, but I was thinking of something like a pig's spine." Thus, Tory tracks down a dead pig.

2. **The fan.** Kari and Scottie call on a business called House of Fans. "I wonder what they sell here?" laughs Kari, then asks shop attendant, Brian O'Connor, to show them the most common fan most people would have in their house. It is a standard 1.3-metre blade fan with wooden blades which rotate at about 40kmh. Then Scottie asks, "If you wanted to get as close as you could to chopping somebody's head off, what would you recommend?" Brian produces an industrial-style fan, complete with powerful motor and metal blades which rotate at 85kmh.

3. **The dead pig.** Back at base, Tory returns. "All right, let's see it!" yells Kari ... who will regret it, being a vegetarian. Tory gets out of the black car (which sets a gothic scene for what follows): "I hope you guys are hungry, I've got 150 pounds of pork!" He shakes the metre-long neck bone. Kari is horrified, on cue: "I didn't think we'd get a carcass with the meat on it, it's just disgusting."

4. **Head-making.** From a previous episode, where gelatin moulds of Adam's head were used to test the myth that tongue studs attract lightning, they have the moulds. So it'll be Adam's blue head on the chopping block in which Tory must first make a hole for the pig spine. "The dummy's head will have to be as close to the real thing as possible," says Tory. "It's a gruesome process." Tory pulls the meat out of plastic, Scottie picks it

up and runs it through the band-saw. She explains: "I'm going to cut this spine to fit in our mould." Kari leaves the room. Adam drops by and inspects the job.

"We're going to need some human cranium."

"We have human cranium?!" exclaims Scottie.

"I have human cranium in storage, yes," he affirms. "You would!"

While the others plug the dummy's internal structure, Kari busies herself with making moulds in the vacu-form machine. Next, they put the moulds upside-down, fill them with ballistics gel, cranium and the spine. Wait, there's more: using his special effects know-how, Tory makes latex arteries which are filled with fake blood. Pleased with himself he says, "Once the blades cut into the artery, blood is going to spray all over the room. It kinda reminds me of *Alien*, when the thing pops out of the womb."

"That is one of the grossest things I think I have ever witnessed," says Kari.

5. **The rig.** Next they turn their attention to the problem of the rigs that will deliver the heads into the fans. Tory asks Jamie, who says, "I've got a thing that we built for a commercial some time ago. It was made to lift a little kid into frame on camera. It'd be perfect." Jamie's scissor-jack saves Tory a load of time and effort. Scottie manufacturers a track and dolly to carry the neck head first into the blades.

THE EXPERIMENT

1. Fracturing a cranium with a common household fan. With the head in place, Tory checks that the scissor-jack is positioned at the correct level. Bulletproof plastic is fitted to protect the team – Tory, Kari, Scottie, Adam and Jamie – while they observe the experiments from outside the test chamber. The stage is set, the dummy unveiled. They build it up on a box. Okay: seal the chamber. The fan spins up to top speed which the tachometer reads at 40kmh. The head is in place for Myth No 1.

"We're going to need some human cranium ... I have a cranium in storage." – Adam

Everybody ready … Tory pulls it. Adam's blue gel-head moves into the ceiling fan. "Ohhh!" The head is rocked, but it doesn't roll. Unsurprisingly *(if you've ever put your hand up into a ceiling fan)* the household fan did virtually nothing to the dummy head. Kari says, "It didn't have much of an impact, it didn't get through the ballistics gel, it didn't crack the skull." Jamie adds, "That actually surprises me because this stuff is not quite as tough as muscle but it certainly held together for this."

It's killin' time. The pig is about to become a little more spineless.

Verdict: The household fan spins too fast for the head to reach neck level, but too slow to do any real damage. The dummy takes the blows on the top of the head and the blades just bounce right off. The "leaping child" myth is busted.

2. Decapitating a head with a common household fan. To re-create the Lover's Leap myth, the head now has to travel along the track and hit the fan right at neck level. The head is in place, the rope around the pulley going to the control panel **… 3 … 2 … 1** Jamie pulls the rope, sending this neck straight into the fan: "Oooh!" It breaks the fan, the dummy gets a sore throat but keeps its head.

Verdict: When it comes to a household fan, both myths are busted.

3. Fracturing a cranium with an industrial fan. It's time to step thing up a gear: bring on the industrial fan. A lot more power, with fast blades like little knives. In fact it goes 85kmh, twice as fast as the household fan.

3 … 2 … 1 … SLAP!

Verdict: The industrial fan managed a bit more damage, perhaps doing enough to break the skull or cut the neck. "That's at least a concussion, maybe a fracture," says Kari. "It may not kill you but you wouldn't want to have this happen to you."

4. Decapitating a head with an industrial fan. This time the neck of the dummy really takes a hit from the high speed metal blades. Next in line, the big one – hold onto your hats – we're going neck first.

5 ... 4 ... 3 ... 2 ... 1 ... Scottie pulls the string.

CRRRACK! The team's special effects expertise is paying dividends. The blades bend on impact. Says Scottie, "The fan actually sliced through his neck, through the jugular and even into the vertebrae. That sounds deadly to me."

Verdict: Deathly, but still not decapitation. Myth still busted. This is the most powerful fan you can buy.

5. Decapitating a head with a fan strong enough to do the job. The plan is to suspend a lawn mower engine from the ceiling, attach large razor sharp blades and let it rip. This should be a real close shave. Calling on the job, Adam sees his own head and adds his glasses. To set the scene, the Build Team then texta-graffiti the wall with the skull and crossbones and the word "Fan of Death". Tory starts up the fan.

"Let's just sharpen these up a little ..."

Since they're heavy blades it takes some turning over. Adam complains he can't see a thing without his glasses, and calls out, "Are you guys ready?"

And **3 ... 2 ... 1 ...** and pulls. **SCREAMS.** Omigod, it's horrible. It's a horror show here. Glasses smashed. Sackcloth, splattered gel. The blade of death annihilated the dummy head, ripping the head/neck to shreds, though not actually doing enough damage to decapitate. The head is dead, but the spine is a still intact.

Verdict: Says Kari: "It didn't decapitate him but it's definitely a deadly blow."

CONCLUSION

Says Adam, picking up the image of his own head, *"This one goes into the Mythbusters Hall of Fame, this is probably one of the most upsettingly lethal things you guys have ever built."*

Scottie: *"I can't believe how hard it is to chop off a head."*

Adam: *"It is apparently very difficult to chop a head off. The injury was definitely lethal, but it still didn't behead our guy. I think that's busted – unfortunately."*

Scottie: *"It's busted!"*

They chopped, and they chopped, but the gel man kept his head. Beaten but not bowed, the Build Team ponder its next conquest.

M= MASS
V = VELOCITY (FEET PER SECOND)
MAIN VARIABLE
MASS

Hold on Tight

Chapter 13

CAN A CHILD HOLDING ONTO BALLOONS VANISH INTO THIN AIR?

The Myth: ***A child can be floated into the air by holding onto a big bunch of balloons***

The Mythbusters are no strangers to unpowered flight. When testing the myth of the Lawn Chair Balloonist (a man called Larry Walters, the real *Danny Deckchair*, who claimed to have flown at nearly 5000 metres) Adam attached his chair to 55 helium-filled balloons, then took to the air and took his life in his hands.

The Mythbusters concluded the myth was plausible. "Larry definitely could have done it," says Jamie. And after himself flying at 300 metres, Adam agrees. So they know you can fly beneath weather balloons if you have enough lift.

But could a bunch of party balloons really carry a kid away from a carnival, as we saw in the *Mr Bean* movie? The Mythbuster's builders are on the case.

'If Maddy were to really get carried away by balloons, where would she end up?" – Scottie Chapman

THE SET-UP

First up, Scottie, Tory and Kari do the maths to find out how large a bunch of balloons they will require, starting with a measurement of how much weight 10 balloons of the standard 28cm arty variety can lift.

- **Calculations.** To estimate the approximate number of balloons required for a total lift, the Build Team fixes easy-to-attach paperclips *(which they weigh)* to 10 balloons. The 10 balloons lift about 100gms of paperclips, hence about 100 balloons will drive a kilogram aloft.

- **Stunt kid.** How much does a child weigh? Weight is an issue here – in terms of blowing up of balloons – and not wanting to turn into a 'blow hard', Tory suggests they fly 'a really skinny kid'. The Mythbusters soundman volunteers his daughter Maddy. She weighs 20kg and reckons she isn't afraid of heights. Again the Build Team does the maths and calculates that it will take something like 2070 balloons to send Maddy, plus her harness, on her maiden flight.

- **Thousands of balloons.** The Mythbusters need more than 2000 balloons, a stack of helium and a speed lesson in tying. So they go shopping at the SF Party store where salesperson Feliciano shows them how to blow up balloons quickly and efficiently. Feliciano reveals that the secret of making really big bunches of balloons is to create lots and lots of columns. "He saved us a tonne of time," says Tory. "I originally imagined each balloon having a string going down to a single point but now we're going to make columns of balloons and tie those together."

Next, the Build Team need to figure out how long it takes to inflate 2000 balloons? Could this task be completed in one day? A time-trial is the only way to find out. The team works on 25 balloons, with three sets of hands. Get set, go!

Tory times it at two minutes, 20 seconds per 25 balloons and passes the data to Scottie for her final word. Kari is optimistic, but they face almost four hours of balloon-blowing time – and they're conscious that they have only got one day to pull this off.

Next, they need a location. Where better to test the fairground balloon myth than an aircraft hangar? So the Build Team shift operations to this appropriate location and start work straight away. Says Kari, "We don't have much time, we have to blow up 2000 balloons and get a four-year-old off the ground by the end of today!"

While the Build Team is blowing up helium-filled balloons and attaching them until their fingers ache, little Maddy is blowing bubbles of her own. She's having a great day on the set, drawing and enjoying lots of attention.

THE EXPERIMENT

TEST ONE

The hangar is ideal, because its roof is high enough for balloon columns – if they work – to lift Maddy without floating her into the stratosphere. They don't want to pick her up somewhere in Alaska, but the question stands: ***where do randomly flown balloons go?*** Scottie explains, "We were wondering if Maddy were to really get carried away by balloons, where would she actually end up?"

The first part of the test involves a simple operation, Scottie and Kari have attached notes to three small bunches of balloons with a request that whoever finds them, please call Mythbusters and state the location.

Outside the hangar, Scottie says, "Ready Maddy, let it go. *Weee!*" Scottie, Kari and Maddy release one bunch each. And off they float. If the power of wishful thinking counts for anything, there'll be news of those balloons before the end of the show.

Hold on to your dreams ...

1900 Balloons, Not Nearly Enough!

Meanwhile, the team looks like it's getting ready for the world's biggest carnival. Kari eventually announces they've tied 1900 balloons ahead of schedule. It's time for a reality check. This project is physically exhausting. Scottie, Kari and Tory are really feeling it and note that the pull of the columns feels as if their arms are being lifted out of their sockets! Nearly 2000 balloons is surely enough!

To avoid undo wear and tear on Maddy, the Build Team test their result-so-far on a stunt kid made of sandbags, the exact weight of Maddy. They hook Sandbag Maddy onto the balloons then weigh her, which is disappointing. Kari can't believe it, "We figured it'd have more lift than that! Are you sure we counted the right number of balloons?" All their high hopes have been slapped down by gravity; they still have to lift another nine kilograms. Never mind the maths, the 1900 balloons only cancel out about half the dummy's weight. The Build Team estimated that around 2070 balloons would be enough to get Maddy airborne, now they need nearly double those numbers, plus they've got to work faster. Not only must they struggle to blow up many more balloons today, but with every minute that passes the balloons that are inflated lose some of their helium and lifting power. Suddenly everything is going wrong ... balloons are popping, balloons deflating, balloons escaping. Still, it's fun watching Tory struggle to retrieve a big bunch from the roof. "We might need about 6000 balloons by the time we get the kid in the air," Scottie despairs.

The Mythterns blow and tie and curse their way towards a balloon bunch of mythic proportions. Time passes, and if their count is correct they are now at 3000 balloons (about 1000 more than they originally expected) and they decide to test the weight once more with Sandbag Maddy. They conclude 3000 balloons will lift almost 17kgs.

"500 more balloons coming up," says Tory.

What credible balloonist would create a bunch this big then hand it to a child? Because of the impracticality of handing out thousands of balloons in a single bunch, this myth is starting to feel busted.

TEST TWO

The Build Team now has 3500 balloons and Maddy's moment has finally arrived.

"Maddy, it's go-time," says Tory, and the Build Team strap the little girl into the harness with safety belts, secure the safety line, check that Maddy is ready and they release her inside the hangar. There is no danger that Maddy will fly off over the horizon because the balloon columns are confined within the safety of the roof.

Result: Maddy lifts five feet from the ground.

RESULT:

HOLD ON TIGHT

MYTH: BUSTED

CONCLUSION:

The Build Team has shown that while you can lift a kid with party balloons, you need far too many for this myth to be plausible. Says Tory: "There is no way that a circus would have that many balloons in one bunch."

"I would think we popped that myth," adds Scotty conclusively.

But remember those three balloons set free to roam the world at the start of this experiment? According to the response from the one (of three) bunches sent off by Kari, Scottie and Maddy, if Maddy had been attached to that bunch, she could have floated 500km away – to the Sequoia National Park!

Well, Maddy had fun anyway.

THE BUILD TEAM

"Public demand means we've had to double our output, and to achieve that we needed to employ another team on board to up the volume of material we were making. It still takes us a year-and-a-half to produce a 26-part series. Right now I'm pushing for a third team but I don't think I'll get it." **- Peter Rees**

M= MASS (SLUGS
V = VELOCITY (FEET PER SECOND)
MAIN
T= TIME
AMOUNT OF TIME FOR DEFLECTION
MASS
MAIN
DEFLEC

Hair-razing?

EXPLODING HAIR CREAM

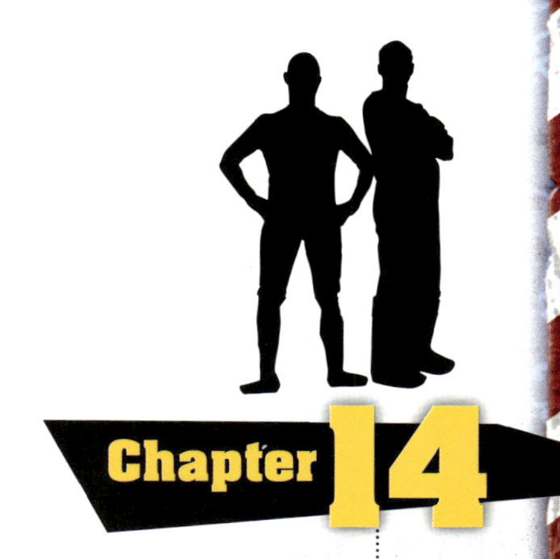

The Myth: ***If sparked, hair cream in an oxygen-rich environment will explode.***

Adam recalls, “In the early 1990s, a pilot getting into his jet on an aircraft carrier put on his oxygen mask and, apparently, as the hair cream turned runny the oxygen blew his helmet off his head and gave him third degree burns.”

“The most famous one of them all of course was the Apollo 1 incident,” says Jamie.

On January 27, 1967, Apollo astronauts Roger Chaffee, Edward White and Virgil Grissam lost their lives in a tragic flash fire aboard their grounded space capsule, which would have been the first manned Apollo mission. A fire swept through the Command Module which killed all three astronauts.

“Faulty wiring and the pure oxygen environment caused a flash fire. Unfortunately, due to the fact that the cabin was pressurised and the door opened inward, the astronauts were unable to escape,” adds Jamie.

For the Exploding Hair Cream Myth the Build Team is getting more and more involved, because Discovery wants 36 episodes per year and the concept is greater than Adam and Jamie’s time constraints.

Says Peter Rees: "Hair Cream is one of my personal favourites, I really think that was a great story. The fact is that you can actually blow yourself up, and I believe it has happened – over Alameda there was an aircraft crash in the early 70s, it was a US Navy Corsair, and apparently the plane crashed into a block of flats and killed five people, and the evidence suggests that the pilot was dying for a smoke on a long flight, took off his oxygen mouthpiece and lit up a ciggie and went boom.

"In an oxygen rich environment, in a high altitude aircraft, it's the entire helmet on your head. In the case of Apollo One, it was the entire capsule. Those guys were incinerated in 30 seconds and I don't think they were wearing hair cream or lip balm. It doesn't matter what the substance is, it can actually be quite explosive. Although we didn't put this in the story we found out subsequently that if we'd added aluminium powder or Vaseline it makes it explosive as well as a fire hazard."

Kari explains the myth, sent by a viewer to the Discovery website: "It seems in the 1950s the Canadian Air Force had a series of freak accidents with their fighter planes. The pilots were supposedly decapitated by an explosion before the planes had hit the ground. It seems a combination of hair cream, an oxygen-rich environment and a spark from the communication system might have made their head explode into flames."

So could the Canadian's hair care regime, plus pure oxygen to his face mask and a simple communication short in their F1-04 Starfighter really lead to a decapitation?

Full head of hair intact, Adam spruiks the product in question.

"I would say that every myth we do has an element of truth to it." – Adam

THE SET-UP

“We know that given the right fuels oxygen can cause a very aggressive fire or even an explosion,” Jamie explains to Kari and Adam.

“It seems to me that the heart of this myth is whether hair cream (and hair) are one of those explosive fuels.

And so, before we start to deal with massive quantities of hair cream and potentially all dying, we need to start out with a scale miniature and see what happens.”

So before launching into a series of potentially dangerous explosions, the Build Team test the plausibility of the myth by carrying out a small scale flammability trial with test-rigs in small explosion chamber.

The need for speed ...

Small-scale test 1: In an oxygen-rich environment, will a short circuit spark ignite hair without hair cream? **Result:** Spark only, a little bit of smoke.

Small-scale test 2: Under the same conditions, will hair rubbed with hair cream go up in flames? **Result:** A fireball.

Now to prepare for the full scale test, which Adam confirms, “All you need to do now is bump it to the full-scale to see if you can get a full-scale explosion and/or decapitation.” The mission, as outlined by Jamie, is to recreate the conditions in a fighter cockpit at altitude.

To re-create cabin conditions, Tory visits the Hiller Aviation Museum, where he finds the model of a US-A4 and takes measurements and other data because a US-A4 has an almost identical cockpit set-up to the Canadian F1-04 Starfighter.

BUILDING A FIGHTER PLANE

Now Tory needs to work out how to build his own cockpit from junk lying around the yard. Fortunately, they have just the thing in the workshop. It's called the Sharammer, a 1350kg length of pipe used as a 'shark ram' to ram a boat in a previous television program. It's big and round enough to make a dummy F1-04 cockpit.

CABIN CONDITIONS

Meanwhile, Grant focuses on re-creating F1-04 cabin conditions at just over 9000 metres. He rigs inlets for oxygen to the pilot and the vessel will be pressurised at 5psi, the same volume as an F1-04 cockpit.

CANADIAN PILOT CRASH TEST DUMMY

Using real hair from a local barber shop, Kari makes some toupees for the Canadian pilot flambé which they play around with. Adam it tries on, and Tory tells Kari, "If Mythbusting doesn't work out for you, you can go into the toupée-making business." She places the toupée on the ballistics gel head, and thrusts it into Tory's face, "Kiss me …!" Next, a Starfighter pilot's face mask and helmet, complete with the eagle motif on the badge.

IGNITION SOURCE

Grant believes the most likely ignition source would have been a short circuit from the 24-volt intercom, so he rigs his ignition system from the small-scale test to the full-face Russian neck helmet. After checking that the seal is air-tight, the Build Team is satisfied that they are ready for action.

Mini Myths

Going down with the ship. *The myth is that the suction caused by the sinking of the Titanic would have sucked hundreds of passengers down with the ship. The Mythbusters sink a nine-tonne boat. Dressed in scuba gear and swimming directly above the boat Mythbusters Jamie and Adam were not dragged down.*

With the hard work already done, Adam arrives and is introduces to the pilot. "Would you like to meet the pilot – it's you!" laughs Kari, showing off the ballistics gel-head. Naturally, Adam tries on the brown wig.

Next, Tory seals the open end of the cockpit and the team runs through the test protocol. The cockpit is pressurised up to 5psi while the rest of the cabin is pressurised with regular air. Grant takes care of the ignition system to simulate an intercom short circuit. Kari goes heavy on the hair cream: "He could be a greaser!" All right. Seal it up boys! We're going up 1.5lbs. Grant starts the oxygen supply to the face mask, and the back of the cockpit blows out. "Well, we got an explosion," says Tory. "This is not the explosion we were looking for, but we got one."

Uncertain about how the cabin will stand up to the pressure, the team keeps its distance.

The first exploding hair cream test ended in near-disaster. With that potential for explosive decompression, safety protocol dictates that this test will have to be reviewed.

THE EXPERIMENT

This calls for a supervising adult. In Mythbusters, this can only be Jamie. "My impression at this point is that it's anybody's guess whether this is going to continue to work," he says. "If I were to bet on it I'd say we're going to continue to have either leaks or further explosions with this."

Jamie lays down the law and the team gets busy. Tory re-plugs the blown cockpit and all system and triggers are extinguished including the fire extinguisher that injects C02, quelling any cockpit fire.

Take 1: Kari initiates the oxygen flow into the mask, and after five minutes Grant initiates a spark from the intercom; there is no ignition from the short circuit. Using the backup from the model rocket igniter, there is smoke coming out of the mask. Tory initiates the fire extinguisher which inadvertently doubles the cockpits' depressurising to a dangerous level. "It's amazing," says Adam. "We got a result out of that. I'm totally impressed."

Tory says, "I didn't think anything was going to happen, I'm really excited we got fire. And then when I hit the C02 to put out the fire the pressure built and we started leaking air in it and the psi of the whole thing went up to 10psi."

Result: The hair ignited, causing the ballistics gel head to melt, however there was no explosion. Under accurate cockpit conditions the Mythbusters did get a pilot hot under the helmet. It seems this myth could be plausible?

Take 2: Kari adds Jamie's facial hair to Adam's toupée and they utilise the airtight MIG helmet. The theory? The next seal will contain the oxygen and create a bigger bang, so this time they don't pressurise it because the helmet is sealed. The first 24-volt short is initiated where the moustache is.

Result: Flames quickly engulf the helmet but there is no explosion – the Canadian kept his head.

Take 3: Trying to create a big explosion, they use only hair care products to try to facilitate an explosive decapitation. Since a lot of hair care products have propellants, solvents and evaporants, that are flammable or even petroleum-based, the Build Team expect quite a big boom.

Result: Ignition! The pressure seal was broken and the windscreen blew out.

"I didn't think anything was going to happen, I'm really excited we got fire. And then when I hit the C02 to put out the fire the pressure built and we started leaking air in it and the psi of the whole thing went up to 10psi." – Tory.

RESULT:
HAIR-RAZING?

CONCLUSION

Back at the workshop **Adam** tells Grant and Kari, *"You guys did some good work on this myth, I really think it's solid. What conclusions did you come to?"*

Grant: *"Amazingly, part of this myth was possible. Hair cream in an oxygen-rich environment (just like you find in a cockpit) and a spark causes fire, not just fire, but aggressive fire.*

Kari: *"But by the specifications of this myth, which included decapitation, we did not by any means achieve that, so I would call it busted."*

Adam: *"I would say like every myth we do there's an element of truth to it but it didn't meet the circumstances, so it's busted."*

Grant: *"Busted."*

Even though things got a little heated in there, our test pilot kept his head.

Don't try this at Home

360 DEGREE SWING SET

The Myth: ***If you swing hard enough you can do a complete 360 degree rotation on a swing set.***

Peter Rees outlines a core Mythbusters credo: "There is obviously a perverse pleasure in watching people suffer. I don't know what it is about Tory and Adam, but Adam in particular is just a disaster area. The number of times you see him on the show swearing because he's jammed his finger, and he does it all the time! He gets caught in the door, he gets bits of carbon fibre stuck in his feet. Jamie says this about him – they're opposites – Jamie says 'Think first then act.' Adam's maxim is 'Act then think' and so these situations are hilarious where he's only got himself to blame.

"When Scottie and Kari were co-hosts, they used to goad Tory on to do things. Kari especially got endless hate mail. 'We know you were setting Tory up for that, you set Adam up …'. She'd always get the blame for some reason. The truth is it's usually me who's set it up."

"Well, that's our job ... to strap rockets to everything." – Adam.

As for the swing set myth, the Mythbusters website has had many, many responses from viewers who believe the 360 degree swing is possible, with many claiming to have done it. "I go on the website all the time," says Peter. "If you can provide us with footage or some piece of evidence which proves it is correct, we will run it on the show. And so far with the 360 degree swing story I do not have a piece of footage, a single photograph, a single article, anything that can definitively say anyone has ever done it, except on one of those hard-arm swings." Says Grant, "The overall plan is to purchase a standard regulation swing set and try it with one of us."

Adam: *"And when that doesn't work?"*

Grant: *"And when that doesn't work, we'll goose it up and take it the next step."*

Yet another common everyday item turns deadly.

The Build Team members are going back to their schooldays to test a classic kiddie myth – it's a dangerous challenge, so leave this kind of mythbusting to the experts. They are going to try to swing 360 degrees and, what's more, today Tory is in charge of safety.

1. **The swing-set:** Tory has already made an alarming find; he reads the disclaimer aloud: 'Warning, this swing was designed for use by children only, not for adult use.' This is not rated for that weight and we have three adults who are about to try this out. So just to be sure Tory adds more mass to find out exactly how safe they are.

2. **Calibrated scale.** Meanwhile, Kari and Grant build a giant calibrated scale which, when coupled with a high speed camera, will allow them to work out whether they are coming generating sufficient force needed for a chain-straight 360 degrees.

3. **High Speed Camera:** The high speed camera will reveal the height at which the chains go loose and the exact velocity of the swinger. This myth is a battle between the strength of the pusher driving the swing up and the force of gravity

pulling the frustrated swinger down. As the swing gets higher its centrifugal force becomes less than the force of gravity, at which point the chain will go slack and the swinger will fall out. If no swingers get close the team can use these figures to learn what it will take to go the full circle.

THE EXPERIMENT

There will be six series of tests:

1. On the swing set using Kari, Tory and Grant
2. On the swing set using a dummy
3. Scale model swing and dummy
4. On a rigid arm circus swing
5. On a swing set using mini rocket power force
6. On a swing set using rocket power force.

1 ON A SWING SET:

1. *Kari.* Even at her maximum Kari is way off a 360. The high speed camera helps the team work out that Kari reached a max-height of 3.3 metres and swung at 7.1 metres per second.

2. *Tory.* Tory breaks the swing and crashes to the ground. So the swing really wasn't rated for adult use. As Tory just couldn't generate enough force to overcome gravity, the higher he goes the harder he falls. But if it's any consolation they got the numbers they needed – 6.8 metres per second. Less than Kari.

3. *Grant.* Before Grant's turn, Tory welds the swing firmly shut. That's not the only precaution because after Tory's accident Grant is suiting up for safety.

Even with two people pushing Grant goes no closer to 360 than Kari and Tory did on their own. Just one 7 metres per second and 3.4 metres in height.

2 CRASH TEST DUMMY ON A SWING:

Kari advances the next plan, which is to use their dummy, Simulaid Suzy, who weighs 29kgs, the average weight of a 7-8 year old. Of course, Suzy can't hold on for herself so the Build Team wire her in. But using Suzy isn't the only change – the team are also halving the length of chain because a shorter chain needs less velocity to make the 360.

1. The Build Team boys heave hard but still no 360.

2. Tory calls in some beefy bikers and two members of the Hell's Angels biker club arrive. Ready, **3 … 2 … 1 …** here we go … **SUCCESS!** Suzy is over the bar, this myth is still alive and kicking. However, although they got Suzy over the top, moments after Suzy goes over the bar she just falls out of the sky. It's not a chain straight 360 so it's back to the drawing board. Says Grant, "Yes, we cleared the bar but it wasn't a full 360, it was more like a 300 then she fell."

Kari: *"Well we know this myth is busted, that you can't under your own power do a 360, but as Mythbusters I want to see what it takes to actually do it."*

3 MINI MODEL TESTS:

Next, the Build Team build a scale model to figure out if it is even possible to do a full 360 degree. And they've started their investigation in miniature, building a swing and a Suzy one-sixth of the original size.

1. *Test No 1:* Tory power on the model. Her speed is 10.9 metres per second, way faster than even Grant swung at, and maybe too fast to replicate on the full size swing.

2. *Test No 2:* There is a way to make it easier – use rigid arms. With rigid arms no energy is wasted in keeping the chain straight, all the forces are channelled into velocity so a much slower speed is needed to go 360°. The rigid arm only needed a velocity of 3 metres per second.

4 TESTS ON A RIGID ARM CIRCUS SWING

The Build Team drive to Trapese Arts in Oakland which has a swing that's guaranteed to do a 360. Says trapeze artist Eric Braun: "I've been in the circus my whole life, I'm a seventh generation circus performer so I really didn't have swing sets. I had Russian swings, trampolines, flying trapezes."

Grant's up first and he's ready for some negative Gs. Grant: "Eric, are you sure I'm safe?" "Er, yes," says Eric indifferently. And with those words of encouragement it's time to swing this thing around.

1. *Grant.* With Eric on board the swing begins to creep higher and higher. They're already over 220 degrees but Grant's not looking so hot. Nausea kicks in at 230. As Grant says he needs a break Tory adds, "He looks like he's about to hurl."

2. *Kari.* Kari is certainly not as green as Grant but after a while she, too, has had enough. Kari is the highest so far, reaching 240. Can Tory make the elusive 360?

3. *Tory 1.* "The challenge is the fact that both of them didn't make it – I have to make it." As they get higher and higher, Tory is beginning to wobble, "I need a break." So close, Tory has got up the highest yet - 280 before his legs went limp. But for this competitor it ain't over yet.

4. *Tory 2:* After a drink of water he climbs aboard once more, and with Kari shouting, "All right, Bellici!" he goes for it one more time. Eric joins him and pushes it higher – already they're past Grant's best efforts – they're over 280 with one final push. But it's just too much for Tory. His mind wants to continue but his body's given up.

5. *Eric Braun, circus performer:* "Well, we came here for a 360, we've got to leave with one," says Grant. "I'll do a 360 for you guys, no problem," smiles Eric, which he does with professional ease. There it is! (cheers).

Result: For rigid arm swings this myth is confirmed – a 360 is really possible.

Time out to consider how to goose up the swing set a little.

5 ON A SWING SET USING MINI-ROCKET POWER FORCE

In classic Mythbusters fashion, the Build Team are going to do whatever it takes to get a chain swing up and over the bar. And they're going to start their extreme experiment with a scale model; strap a few rockets on mini-Simulate Suzy and just watch her go round the swing set.

1. Test 1: The little rocketeer went over the bar, but not quite as planned.

2. For Test 2 the rocket is positioned exactly horizontal for maximum velocity. The horizontal rocket sends Suzy into a cataclysmic spin and blows off her hand.

3. So for Test 3 the rocket is positioned vertically for maximum tension on the chain. The vertical rocket was pretty useless, it's going to be a combination that's needed for this myth ... something like, say, 40 degrees.

4. Test No 4: The rocket is 40° to the horizontal. This was spot on: Suzie goes 360 degrees and beyond, so the next stage is to replicate this result on the full-size swing by strapping on some mammoth missiles.

6 ON A SWING SET USING ROCKET POWER FORCE

Now at Hamilton Air Force base, the Mythbusters scale things up and strap rockets onto their full-size swing to replicate this feat. Luckily they've got amateur rocket expert Erik Gates to help them to jet-simulate Suzy 360 around a swing set.

1. Because Suzie is so much bigger than her scaled-down cousin the team decide to attach four rockets. That's 400 lbs of thrust at the magic 40 degrees angle. Once they're on, the igniters are put in and wired up in series. Says Erik, "If one fails the others will work, that's why we wired them up in series." It does a 360 – but only one side of the rockets ignited, which sent Suzy into a gut-wrenching spin.

2. This time Erik puts just one rocket each side – Suzy is ready for another wild ride. So what went wrong this time? The high speed shows that one side fired just before the other. It's a fraction of a second but it's enough to throw Suzy into a crazy twist.

3. They now attach Erik's single biggest rocket dead centre. But because they're worried about Suzy's weight they opt for a launch angle on 60 degrees. Third time unlikely, but it's not all bad news: when the rocket she didn't spin; she stayed in one direction. That means all they need to do is drop the angle and she'll go forward and be pushed down.

4. They drop the angle to 40 degrees so everything matches the scale test.

... 3 ... 2 ... 1 ... and she's over at last. Simulaid Suzy finally goes the distance, and the high speed camera shows the bars were straight the whole way round. Says Kari, "I didn't think it was actually going to work." Adam adds: "I didn't think it was going to work most of the time."

Result: A perfect demonstration of a swing set 360.

RESULT:

DON'T TRY THIS AT HOME

MYTH: BUSTED

CONCLUSION

So in the final analysis what is the final verdict: can you do a full 360 in a swing set – busted, plausible or confirmed?

If at first you don't succeed ... strap rockets to it.

Tory: *"Totally busted, not possible busted."*
Grant: *"Not possible, busted."*
Kari: *"Busted, but the rockets were spectacular."*
Adam: *"Well, that's our job … to strap rockets onto everything."*

"Busted, but the rockets were spectacular." – Kari.

STEAK
PHOSPHORIC ACID
STEAK
COLA

Mystery in a Can

Chapter 16

COLA: WHAT CAN'T IT DO?

The Myth: ***Everyday cola can clean off rust, dissolve a steak and a tooth, and clean the contacts on your car battery.***

Cola. America's favourite drink is a carbonated soft drink containing an extract of the cola nut. This drink has always attracted its share of popular myths, some of them true, most of them not, but all of them challenges for the Mythbusters.

Adam: *"Jamie, how many things can we do with cola? Some of them sound like they're pretty reasonable, and some of them sound as if they're absolutely ridiculous."*

Jamie: *"Lots – we can clean up blood."* ... **Adam:** *"We can dissolve a t-bone steak."* ... **Jamie:** *"We can clean chrome."* ... **Adam:** *"We can clean the contacts on your battery."* ... **Jamie:** *"We can clean a rusted bolt."* ... **Adam:** *"We can clean your greasy laundry."* ... **Jamie:** *"We can dissolve rust."* ... **Adam:** *"Dissolve a tooth."* ... **Jamie:** *"You can clean your car engine."* ... **Adam:** *"It might also be an effective spermicide. ...*

Well, let's go and get some blood, and steaks, and chrome, and rusty bolts and get started."

"It might also be an effective spermicide ..." – Adam.

Myth No. 1: Cola will dissolve blood

With Jamie lying on the ground and Adam chalking a crime scene outline around him, the question is: will cola shift those hard-to-remove blood stains from the Mythbusters 'crime scene?' … which is the bumper bar of the truck that allegedly hit him? While Adam is chalking him up, Jamie surprises Adam, "You realise this white line thing is all a myth – they don't actually do that?"

A puzzled Adam replies, "You're kidding me?"

Animal blood is used to keep the test as authentic as possible. Uncertain about what happens next, Jamie questions, "How long do we have to wait? Is it like the amount of time it would take for an ambulance to get here?"

While marking time, allowing the blood to soak, Adam produces a siren and, with a plug for Mythbusters, he switches it on and says: "For the realism and effectiveness we always strive for, here it comes …" and his siren goes off, light flashing.

After two hours in the California sun, they check their results. Jamie remarks, "It's rejuvenating the blood."

"Isn't that what it's advertised to do?" quips Adam, hosing down the 'crime scene'. And then they check the offending vehicle.

Result: Most of this blood is washed away. The cola helped - not by a whole lot, but it did help. The cola area is cleaner than the other area, and Adam admits he saw more blood streaks on the non-cola bumper.

Myth No 2: Cola Cleans Chrome

Will cola shine the chrome on your car? The Mythbusters are about to compare cola against a leading brand of chrome cleaner. No-one volunteers their car for the cola-cleaning rust test, so Jamie rolls in a junker to see if it can take all the cola they can throw at it.

"It's a new approach, and that's what sold the show – we don't just re-tell the legends, we put them to the test."– Peter Rees

“I didn’t know it had brakes until you stopped,” says Adam. Jamie shrugs, “I didn’t either”. It’s a junker all right.

In white overalls, Jamie rubs the bumper bar with cola while Adam uses a standard cleaning product. After a short observation we get feedback: “I think it ionises the cola particles and brings the rust to the surface,” notes Adam.

Jamie is pleased too: “I think cola has done a great job, it has a foaming scrubbing bubble going on in there, it leaves a sticky mess but that’ll wash off. Your side sucks.”

Adam shucks, “My side sucks, I’ve got to say cola is a fantastic bumper cleaner.”

Result: High marks for shining the chrome.

Myth No 3: Cola Frees Rust

The myth is that cola’s low phosphoric acid content is effective against rust. A cola-soaked rag should be enough, so the Mythbusters apply it, and Adam wants a five-minute break. so Jamie jokes he’s got to drink some cola in the interim.

Add some fizz to your life ...

The experiment is dead. Adam says “It ain’t moving”; Jamie agrees. Then he gets a hold of it but his results have nothing to do with the cola effect. “I don’t think it really penetrated that well,” says Jamie.

Adam: *“I think this one was a total failure in that regard.”*

Result: It maybe removed a little surface rust on it, that’s about it.

Myth No 4: Cola Corrodes Coins

In this test, the cola will be compared to pure phosphoric acid, the active ingredient in cola that gives the drink it’s sharp taste. Most food contains natural occurring acid, much of it stronger than phosphoric, so how will the cola react with coins? The pennies soak for 24 hours.

Flipping and fingering the coin Adam observes, "I see something interesting here Jamie," observes Jamie. "I know for a fact that the penny which was in the cola had a side that was down with an air bubble. We can see that where there was air it didn't clean but where there was air the cola did."

Jamie: *"Wow, that's neat. The cola did a better job than the phosphoric acid."*

Adam: *"I'd say it did a much better job and the phosphoric acid didn't do much at all."*

Result: Cola cleaned this penny really well.

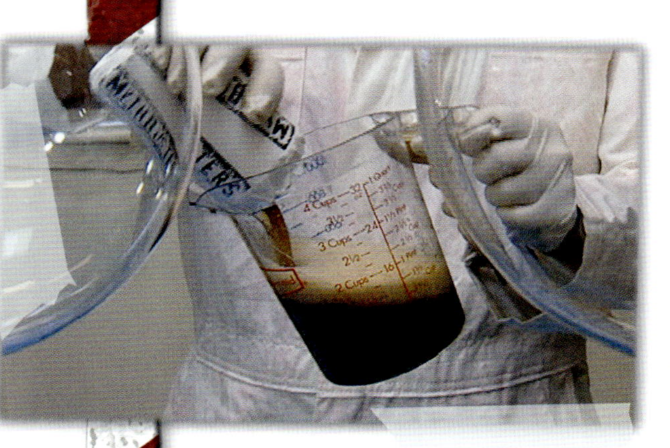

Even dealing with such a broad range of urban myths, the boys retained their highly scientific appproach.

Myth No 5: Cola Corrodes Teeth

There are plenty of legends associated with cola, some of them sound feasible. Everyone's heard about cola's corrosive qualities – can cola dissolve teeth? In their workshop, the Mythbusters set up an experiment where one tooth will soak in cola for 24 hours, the other in pure phosphoric acid. Will they still be there after an overnight soaking?

"They're still here rattling around in there," says Adam. But Jamie adds that the cola made it a yucky brown.

Adam is disgusted and wrongly observes that alternatively the phosphoric acid had no effect. Jamie pulls him up. "Actually," he says, "there's a lot less of that tooth than there was to begin with."

Adam: *"Really?"*

Jamie: *"Yes, it's eaten maybe half of it away. It looks nice and white in comparison but I don't think I'd want to put either of them in my mouth based on that."*

Result: In a 24-hour period cola could never dissolve a tooth.

Myth No 6: Cola Dissolves Steak

With white-gloved hands the Mythbusters drop two steaks in their respective cola / phosphoric acid glass containers, where they will stay for 48 hours or so. Jamie jokes, "That looks yummy."

Adam says, "Now that's a marinade. I'll have to try it – you know my recipe looks really, really tasty." Big mistake; it's Adam who is most disgusted by the results.

Adam observes that the steak tends to float in the acid but not in the cola. Jamie is surprised, he says he would have guessed the other way around. Says Jamie, "Well it's done something in 48 hours."

Adam: *"We've given it a little bit longer than that and you've put your steak in just phosphoric acid."*

"See that?" says Jamie, forking it, watching it fall apart. "It's totally broken it down."

Adam: *"It did, this is the cola steak with a fine glaze of mould. Well there's nothing remotely dissolved about it, I'm trying to speak through my mouth because I don't want to smell this."*

Jamie: *"It doesn't smell bad at all, it smells okay, it smells like 'food'."*

Adam: *"I wouldn't eat it."*

Jamie: *"I wouldn't either."*

Adam: *"So – after all your talking you wouldn't eat it!"*

Result: Myth busted – cola did nothing to a steak.

Myth No 7: Cola Cleans Battery Terminals

Can cola clean battery terminals? To find out, Adam and Jamie track down a well-corroded battery and brush on straight cola on one terminal, and baking soda on the other. They paint it on.

"Personally I don't see any major difference between them," concludes Adam. But Jamie says, "Yes, but the cola actually does work."

Adam reckons the positive results are largely because cola is a liquid, and any other liquid would work just as well. "I'm going to try straight water on this one," he says. "It might be that the cola is simply a liquid, not that it has any special properties, because the water without anything just cleaned up that terminal pretty fine."

"That's true," adds Jamie.

Result: Yes, cola cleans battery terminals, but so does any other liquid.

"It doesn't smell bad at all, it smells okay, it smells like 'food'." – Jamie.

Myth No 8: Cola Cleans Off Grease

Is cola the great grease stain remover? Jamie gets under the bonnet of a car and rolls around. He is already sporting a grease handprint that Adam smeared on his overalls during the earliest cola experiments. Now it's time to find out whether these grease stains can be removed with cola.

One sample is placed in cola; the other in a leading brand of detergent, and both left to soak. Four days later, back in the warehouse checking results Adam says, "For my money unagitated laundry detergent didn't do squat."

Result: Cola turned the material brown but it didn't do anything at all to the grease. "So strike another one off the list," says Jamie. Busted.

You want brown clothes? Wash them in cola ...

Myth No 9: Cola Is A Useful Degreaser

A real cola cleaning challenge: an old greasy engine. Adam applies cola on the left side, Jamie pours the control substance (water) on the right.

"I tell you right now, I wouldn't do this to my car," says Adam. "Should we let it sit 10 minutes or so and then come back?" Jamie agrees.

Result: After a hose-down, as an engine degreaser, it's busted. It cleans up a lot of the dirt but the grease is all still there, same for cola as for the water. It removes a little bit of corrosion but not a whole lot.

Myth No 10: Cola Destroys Car Paint

Will cola spills damage car paint if not cleaned off? Once again the cola is competing against the concentrated power of phosphoric acid. Jamie and Adam pour both on the vehicle. Twenty four hours later, what's the verdict?

Result: The phosphoric acid is bright white but cola is definitely not effective in ruining a paint job. Another one busted.

RESULT:
MYSTERY IN A CAN

Myth 11: Cola Is An Effective Spermicide

The Mythbusters will combine one small sample of sperm with cola and another sample with a control substance. Under the microscope they will estimate the score between alive and dead. This is on-screen sperm.

Adam: *"This like those films in high school!"*

The control substance is two drops of saline solution, which Jamie thinks ought to be kinda easy on the sperm. Here we go … for 60 seconds, Jamie counts live sperm in the saline and claims to have countered 53 live ones in a minute. The 60-second cola count results in 85.

Dr Paul Turec is a male reproduction specialist at the Department of Urology, UCSF. He's heard hundreds of similar claims and commenting on Jamie and Adam's countdown experiment he says, "The experiment shows what I thought would happen, which is cola will serve to dilute the sperm but won't be necessarily toxic to it. A spermicide is different. A spermicide actually stays in there so it doesn't actually dilute anything, it just kills the sperm on contact."

Result: Cola is not an effective spermicide.

CONCLUSION

Adam: *"So Jamie, we've been testing cola for a week, what have we ended up with?"*

Jamie: *"The only thing it really did well was it did a great job of cleaning chrome. Otherwise it sort of cleans up blood, it cleaned up a penny okay, but nothing really* **spectacular.** *But cleaning chrome was great."*

Adam: *"I agree, that's the best thing it did of anything."*

M = MASS (SLUGS
V = VELOCITY (FEET PER SECOND)
MAIN VARIABLE
T = TIME (MAXIMUM AMOUNT OF TIME FOR DEFLECTION)
man
Pascal = Newton/sq. meter
Crack
sq. meter

Chapter 17

What's Cookin'?

THE COMMON KITCHEN MICROWAVE HAS CREATED PLENTY OF FOLKLORE

The Myth: ***You can cook your own intestines by standing too close to the microwave.***

First, a brief lesson: to cook food, a microwave oven uses a device called a 'magnetron' to release short, high-frequency radio waves *(microwaves)* into the oven cavity. These react with the water molecules in the food, causing them to vibrate at enormous speed. The heat occurs when waves penetrate food and set its water, fat, and sugar molecules in motion. That friction creates heat which cooks the food.

Microwaves are a non-ionizing energy, so they can produce thermal energy without making food radioactive. In the early days it was hailed as a revolutionary kitchen convenience. However, the humble microwave oven is a magnet for urban legends, so the Mythbusters put some of the main ones to the test.

"Microwaves are really cool, they're like X-ray guns and the same thing that makes people afraid of them is what's always fascinated me." – Jamie.

Of all the myths surrounding the microwave oven, none is more famous than the poodle sent for a spin in a microwave to dry its fur. So the Mythbusters meet a two-year-old show poodle named Jazzy, who is being groomed for her Mythbusters guest appearance, and has been primped and preened for her moment of truth.

"There are some things even we as Mythbusters can't do, and I think this is one of them," Adam admits. "There's just no way we could put a poodle in a microwave. It might dry her hair but it'd kill the dog. Fortunately for the viewers, this one's just going to have to remain a legend."

It is similar, however, to another myth: millions of Americans tan at indoor salons, and in recent years one particular microwave myth has spread – the myth of the tanning bed cooking the insides of the bride-to-be?

According to folklorist Heather Joseph-Witham, the story is about a woman who is irritated by time limitations at tanning salons. "The woman is getting ready for a special

MICROWAVE OVEN FROM HELL

Apart from testing these microwave myths, Jamie wants to build the microwave oven from hell. So he and Adam visit Friedman's Microwave Ovens, a 25-year-old family-owned retail chain based in Oakland, CA, which has been selling microwave products ever since they came onto the market.

Jamie startles the salesperson by explaining, "The main thing I'm interested in is raw power." Salesman Michael Friedman replies, "Well, the average microwave today is about 1100 watts but you can go up from there and you can go down from there. How many people are going to be using the oven?"

"One very strange television show," laughs Adam.

The microwave retailer explains how the oven works, "This is a little radio transmitter, you plug it in and then it converts its voltage and out comes the microwave." This suits Jamie. He buys four, takes them apart and pulls out the magnetrons from each.

"I'm going to make this super-powerful microwave gun that I can heat things up in my shop with and that to me is a lot of fun," he says. It looks like he's serious about building that super microwave machine.

Finally, Jamie has finished his construction. "What we have here is a super-dooper microwave oven," he says. "I've taken the magnetrons out of four microwaves and I've put them in a box all aiming at a single glass of water, so in theory they should heat up that glass of water about four times as fast as a regular microwave would. We'll find out."

We're ready for a mega-microwave test firing, the Mythbusters take over behind a shield. Jamie flips the switch. The room is filled with an angry buzz. But it doesn't look like anything is boiling. "You've made a refrigerator," laughs Adam. "For all that humming you ended up with bunkus!"

Perhaps Jamie is no evil genius after all.

"There's just no way we could put a poodle in a microwave. It might dry her hair but it'd kill the dog. Fortunately for the viewers, this one's just going to have to remain a legend." – Adam.

occasion, she decides she'll look fabulous in her dress but she's very tanned. So she goes to a tanning salon and she's very annoyed that they have a half-hour daily limit on tanning, so she goes there and also to several other places so that she's tanning for hours a day. After a few days of this her husband says, 'Honey, you smell kind of funny', so she showers, but still smells funny. So she goes to the doctor and, after numerous tests, she finds that she's microwaved her internal organs and only has two weeks to live."

Joseph-Witham adds that this myth has been told and re-told to the point where the industry magazine Tanning Trends went into print to counteract the bad publicity press generated by this myth.

Rick Mattoon, Director of the National Tanning Training Institute, claims to have heard a different version of this story every month for the four years he has been involved in the tanning industry. "I keep an archive of all the different stories that we've got and it's very interesting," he says. "It grows from the local level to the national level, and always has a tendency to go back to the microwaving theory or story."

Can a tanning bed act like a microwave oven? The simplest way would be to strip off and submit to the simulated sunlight, but our boys don't look like the tanning types … which explains why two chickens have been purchased from the local deli.

THE EXPERIMENTS

Myth No 1: A Microwave will cook your insides

The Mythbusters will give the birds four straight sessions on the tanning bed and check the result. After a total of 48 minutes on the tanning lights – to a point where

a human would have severe sunburn – the Mythbusters cut the chicken in half and Adam remarks, "It's pink and perfect, the insides are not cooked at all."

Adam: *"Do tanning bed put out microwaves?"*

Jamie: *"They put out nothing of the sort, they put out light waves and that's all there is to it."*

For the myth to work, microwaves would have to cook from the inside out, and that's easily tested – simply throw a roast in a microwave for 20 minutes and carve

CONCLUSION: The chicken is cold on the inside, proving it is not heated from the inside-out but from the outside-in. Says Jamie: "The myth is totally busted."

Myth No 2: Microwaving metal will cause your oven to explode

The myth that microwaving metal will cause an explosion is well known. Will metal turn your microwave oven into bomb?

Now listen here, Jamie ...

Adam and Jamie are going to microwave several metal objects. Just in case something dramatic does occur the pair watch behind a protective bulletproof shield. It's time for the first metallic test subject.

Spoon: Nothing happened. After the spoon has been in the oven for two minutes. Adam says, "I don't see nothing" so he touches it and burns himself. "The spoon was really hot, but it didn't do anything. It didn't arc, it didn't buzz."

Fork: Maybe the sharp prongs will cause sparks to fly? Jamie observes, "Not a thing."

Scrunched tinfoil: Adam and Jamie have heard that tinfoil scrunched into a ball can have an effect, which it does ... but no explosion. "That's cool," says Adam, "like something out of Star Trek!"

"Beautiful," adds Jamie, because the microwaves cause an electric charge to build up in the metal and crunching the foil allows the charge to jump the gap. Jamie tries again, this time folding the foil into an accordion sharp. Spectacular, yes; explosive, no. Is the metal in a microwave myth true or false? Jamie says, "It's false with provisions, only in the case of an object that is made out of metal is contacting the sides you can create some sparking there that would feed back into the magnetron and reduce it's life span."

MYTH: BUSTED

CONCLUSION: So, it's myth busted.

Myth No 3: Water explodes after being boiled in a microwave oven

It's time now for Adam and Jamie to investigate a microwave myth that is potentially lethal. Does water explode after being boiled in a microwave oven? The myth is that, after heating water in the microwave oven, after you take the cup out of the bowl the water will explode in your face, injuring you or burning you in some horrific way.

The first job is to time how long water takes to bubble. After a little research, Adam and Jamie learn that the only way water could explode is if it is heated beyond boiling point without actually boiling. Tap water always boils because it contains impurities; distilled water has no impurities, that means no boiling over.

So they place two cups of water in the microwave oven – one distilled, one tap – and when the tap water boils the Mythbusters will know the distilled water has become super-heated. Just dropping a sugar cube in the distilled water should then make the boiling process happen instantly and violently.

This they do. There is instant boiling and an explosion of water with Adam exclaiming, "Spectacular!" and Jamie summarising the result: "It erupted, it leapt the container. And that should suffice."

Therefore, if straight tap water can explode how likely is it people would be scalded by exploding beverages?

Adam extrapolates, "On your average day in your average house it's not very likely, but for the hundreds of millions of people using microwaves every day to boil water, I'd say it could happen a few times a year around the world."

CONCLUSION: The myth is absolutely true. If someone were holding the cup they would definitely have severe burns.

MYTH: CONFIRMED